Improving Urban Schools

Equity and Access in K–12 STEM Education for All Students

A volume in
Contemporary Perspectives on Access, Equity, and Achievement
Chance W. Lewis, *Series Editor*

Improving Urban Schools

Equity and Access in K–12 STEM Education for All Students

edited by

Mary Margaret Capraro
Texas A&M University

Robert M. Capraro
Texas A&M University

Chance W. Lewis
The University of North Carolina at Charlotte

IAP

INFORMATION AGE PUBLISHING, INC.
Charlotte, NC • www.infoagepub.com

Library of Congress Cataloging-in-Publication Data

A CIP record for this book is available from the Library of Congress
http://www.loc.gov

ISBN: 978-1-62396-230-2 (Paperback)
978-1-62396-231-9 (Hardcover)
978-1-62396-232-6 (ebook)

Printed and bound by CPI Group (UK) Ltd, Croydon, CR0 4YY

CONTENTS

FOREWORD

Gerald A. Goldin
Rutgers University

When problems facing a society take the form of tangible losses or visible threats, the formulation of responses can be fairly straightforward. Natural disasters such as earthquakes, floods, or epidemics; perceived economic or military challenges from international rivals; incidents of terrorism; or social problems such as unemployment, malnutrition, or high crime rates, are examples of such visible or tangible problems. Faced with such situations, our media highlight the issues, the public becomes aware, and there may be a sense of urgency. Our thinkers debate alternatives, public institutions formulate new policies, and government and the private sector allocate resources to implement them.

But societies have far greater difficulty identifying and responding to the *invisible* losses and the *hidden* dangers. Missed opportunities are not recognized as having been possibilities, and perhaps will never be. We have seen how dangers such as those associated with the safety of nuclear plants, ecological impact and climate change, or the deregulation of banks and currency instability often go unaddressed until actual disaster strikes.

In a certain way, the inequities in STEM education—science, technology, engineering, mathematics—constitute a problem that is visible. Various assessments suggest persistent "achievement gaps" in the United States in school mathematics, associated with poverty, large inner-city educational systems, race, and ethnicity. The long-standing underrepresentation in vari-

Improving Urban Schools, pages vii–ix

ous STEM disciplines and careers of women, as well as African American, Hispanic, and other racial and ethnic minorities, is well known. Today, higher education opportunities and jobs are systematically denied to thousands of undocumented immigrants who have grown up in America. The immediate consequences of such inequity pertain, of course, to the people affected—the educational and economic achievements foregone, the career goals missed, the children experiencing "a dream deferred" more than half a century after Langston Hughes penned his most famous poem.

But the *hidden* consequences of STEM inequities for society are, in my view, far greater. They are, however, unquantifiable, because they have to do with possibilities *not* realized. Who can know what fundamental scientific discoveries were not made by the brilliant women who did not become physicists, chemists, or mathematicians? How can we guess what cures and therapies were not discovered by the talented African Americans or Hispanics who did not become biochemists, mathematical biologists, and medical researchers? Or what new technologies imaginative undocumented immigrants did not develop, and what industries did not grow as a result? How can we measure the value of the influence that a population universally mathematically, scientifically, and technologically well educated did *not* have on public policy; or the social value of all the life decisions that might have been taken with scientific understanding but were not; and the economic choices that might have been made with mathematical understanding, but were not?

That is, from one point of view we may simply say it is *unfair* for children to have fewer educational and career opportunities due to their race, ethnicity, immigrant status, gender, or economic status—an idea conveyed by the term *equity*. But from a different point of view, *our society as a whole has lost*, and continues to lose, all of the social and economic benefits to be derived from a major pool of individual and collective talent. Here may lie some of the most profound consequences of inequity in STEM education, unconsidered and unvalued.

The present book is filled with ideas for improving access to effective STEM education in America's urban schools. These include successful school administrative policies in a STEM-focused school, strategies that have resulted in the success of eighth-grade algebra for all students, new applications of technology, culturally relevant project-based learning, innovative professional development models, and ideas for private-sector collaboration. Some of the ideas discussed have proven themselves, with ample research to document their effectiveness. Others are more speculative, but have reason to offer promise. In addition, the book offers descriptive information pertaining to student engagement, factors influencing the pursuit of mathematics and science by females and underrepresented minorities, and other relevant aspects of the present situation in urban schools.

As the various authors show, there is no absence of proven techniques and no absence of new ideas. None of these are based or need to be based on the current, damaging trend toward punitive measures tied to student test scores. What is needed now is the political will, supported by the knowledge of thoughtful observers and practitioners, to put in place adequate support structures for STEM teachers, embracing schools and the communities they serve, to implement constructive processes, together with a cultural shift, in which quality STEM education for all becomes the societal norm. The costs may be significant, but the benefits are boundless. The goal is urgent, as the generations for whom STEM education is lost cannot be recovered. It is to be hoped that the present volume will help pave the way toward formulating such aspirations and, in the near future, achieving them.

CHAPTER 1

INTRODUCTION TO STEM EQUITY

Robert M. Capraro
Texas A&M University

In the United States today, the funnel leading students to postsecondary science, technology, engineering, and mathematics (STEM) is constricted at best. Well, this is not true; the conduit could be too small but the funnel ginormous. For example, the close approximation for the equation for laminar flow through a conduit (i.e., not turbulent, given a direct path) is

$$\text{FR} = (\pi\,(R^4)\,(P_f - P_o))/(8\,NL).$$

FR is the volumetric flow rate of the liquid (e.g., gal/sec), Pi = 3.14159, R is the radius of the conduit, P_o is the fluid pressure at one end of L, P_f is the fluid pressure at the other end of conduit, N is the fluid's viscosity, and L is the length of the conduit. The temperature dependency here is in N because the fluid's viscosity depends on its temperature. So by the equation, the problem could be in the supply or in conduit between secondary and postsecondary education. It is likely to be a combination of the two, with a confounding for students of color and women.

Improving Urban Schools, pages 1–5

In this book, you are challenged to consider the issues of STEM equity not in the common terms such as gender, ethnicity, or race but also in terms of identity, technology, opportunity, commitment, and challenge. Students in high school who are undocumented face myriad problems, including conflicts in identity between the self as capable and academically competent and the self as needing to earn an immediate living. Undocumented students are tracked like they have the same opportunities for postsecondary education as their documented counterparts, when in reality, if they matriculate to postsecondary institutions, they do not qualify for in-state tuition and often face paying international student fees. Going home is not a course of action, because there is likely no immediate family who still reside in their birth country. So they often bypass postsecondary education for trades wherein there are no birth origin penalties. What does equity look like for them? Advances in technology can provide incredible educational benefits, but for those who are economically challenged, these technologies prove to be barriers. The barriers are not an issue of access but of immediacy. Students who have smartphones have access to the sum of knowledge available on the Internet. A student who does not own this device must still compete with others who have it. This dichotomy creates an artificial chasm in student learning and perceived capability. The lack of opportunity in schools with a disproportionally large economically disadvantaged population also creates a false dichotomy when those students enter college and now have to match the comprehension and learning rates of students who attended more affluent schools when they do not the same background knowledge. Commitment and challenge are interactive and iterative. When students are adequately challenged, it can influence their commitment; and when students have high levels of commitment, they actively seek challenges, even when those challenges are extracurricular. Therefore, as we consider ideas of equity, it is essential to consider these four points for *all* students and remember that students come to both K–12 and postsecondary institutions from many different programs. That variation must be attenuated and carefully shepherded to help students with STEM dreams find them in reality.

In this book, many issues will arise that influence equity in STEM education. We challenge you to examine the issues and determine if the issues result in too narrow a funnel or too small a conduit. Think about issues broadly, not just based on traditional demographics. Consider the infrastructure that facilitates STEM education, from national, state, and local levels, including state-supported STEM centers and education service centers. Think critically: do they facilitate STEM education or the status quo? Certainly the argument that only the best and brightest should be STEM professionals is common; however, we must consider what it means to be bright enough. As a society, we can no longer believe that only the top

10% should go on to prestigious and lucrative careers, because there are others who can also meet expectations when we as an academy fit students to standards and not to a bell curve. Using a bell curve *might* have been useful at one point in the American educational system, but no longer. As a global economy becomes more and more important for a country's economic success, we can no longer have huge volumes of students in courses they are fully qualified to take. The practice of fitting people to a bell curve simply because of a priori decisions about human performance oftentimes leads to 40% of a class receiving a C and 30% receiving a grade worse than a C. Grades should be accurate predictors of success in subsequent coursework. There needs to be greater alignment between and among courses and sections. There must be greater involvement and curriculum articulation among K–12 schools, community colleges, and universities to create a seamless path between them.

The situation in K–12 schools is not one in which students are first and foremost. Unfortunately, teacher accountability has placed the focus directly on the teacher and not the student. For a moment, try to name the programs anywhere in the United States wherein students and/or parents experience consequences for underperformance on state minimal skills tests. You cannot think of any? Well, there are none. The system is a teacher-accountability instrument, so the focus is the teacher. The flaw is believing that the teacher is the sole factor in student learning and success. Now think about the programs in the United States in which teachers potentially lose jobs, are forced to change schools, teach a subject other than what they had been teaching, or are forced out of teaching because of poor student performance. Yes, you probably know of a program in your school system or one affecting a teacher near and dear to you. The focus must be the student, individual success must be a priority, and the federal and state governments must put dollars into supporting programs that place the commitment partially in the hands of students and their families. Vouchers and charter schools are a simplistic and unrealistic solution to improving STEM education. Choice simply allows high quality students to move to schools away from children and parents who lack commitment. Concentrating the negative attitude toward education makes vouchers and schools of choice appear to outperform their counterparts. Honest and hardworking educators know that summer programs (before- and after-school programs) are one means to effect change. Requiring students and parents to participate when students underperform will place an important share of the responsibility for the educational system where it belongs and will likely result in greater equity in STEM education. K–12 teachers must be empowered to take curricular risks and to reinstate creativity and problem solving along with a great variety of assessment types that will take student diversity into account.

It is with great enthusiasm that I anticipate the rest of the chapters in this book. I believe they will provide insights into issues surrounding equity in STEM education and suggest a research agenda for the next decade.

This volume is organized into 11 chapters. In Chapter 2, M. Suzanne Franco and Nimisha H. Patel examine student engagement and college readiness among students attending STEM schools. The authors also provide information about career exploration tools for students. In Chapter 3, Robert M. Capraro, Mary Margaret Capraro, and Michael Muzheve explore how collaboration between private-sector entities and K–12 schools can be used to advance STEM education initiatives. The authors provide practical recommendations for developing and maintaining successful partnerships. In Chapter 4, Jamaal Young, Jemimah Young, and Christina Hamilton describe a framework for how STEM Project-Based Learning can be made more culturally relevant and thus more meaningful for students from diverse backgrounds. The chapter provides recommendations on how teachers can implement culturally relevant STEM PBL within their own classrooms. In Chapter 5, Frances Spielhagen documents the implementation of mathematics policy reform in one school district to increase mathematics achievement among underserved students. This chapter provides insight into the situation that prompted reform and the lessons learned along the way. In Chapter 6, Brian Boyd and Matthew Grushon examine the experiences of a successful STEM-focused school within its first 3 years of existence. The authors focus specifically on the administrative structure and practices that contributed to the school's early success and provide recommendations for how these practices can be enacted in other school settings. In Chapter 7, Eduardo Mosqueda and Saúl I. Maldonado present research examining the relationship between racial and linguistic diversity and participation in advanced mathematics courses. The chapter closes with recommendations for increasing participation in advanced mathematics courses among underrepresented students. In Chapter 8, Trina Davis explores how 3-D virtual environments can be used in preservice teacher education to advance STEM teaching and learning. The author illustrates the use of 3-D environments through her description of *Second Life*, a virtual middle school classroom. In Chapter 9, J. Eli Crow, Teresa J. Kennedy, Michael R. L. Odell, John D. Ophus, and Jason T. Abbit describe a professional development model aimed at increasing teachers' use of learning technologies during instruction. The authors show how this promising model can be used to close achievement gaps among underrepresented students and majority students. In Chapter 10, Meredith Jones examines the persistence of females in the school-to-career STEM pipeline and the different factors that contribute to a gender gap in STEM education and careers. This chapter closes with recommendations for increasing gender equity in STEM education In Chapter 11, Brittni D. Jones and William F. Tate examine equity in science education within a metropolitan area. Based on their findings, the authors provide programmatic and policy recommendations for increasing science proficiency among all students.

CHAPTER 2

STEM SCHOOLS

Facilitating Student Engagement and College Readiness

M. Suzanne Franco and Nimisha H. Patel

For more than 60 years, the United States has continually made efforts to transform its public education system with the goal of improving students' academic achievement and increasing graduation rates. Reform examples include topics such as ability-tracking students and the development of content standards (Lee & Ready, 2009). More recently, the emphasis has been on the development of new teacher-evaluation systems (Darling-Hammond, Newton, & Wei, 2010) and increasing the number of individuals enrolling in higher education (Race to the Top, 2011). Meeting the Race to the Top goal requires an increase in academic rigor within high schools (Kuo, 2010). The development and implementation of science, technology, engineering, and mathematics (STEM) education is meant to address the need for increased rigor while facilitating careers in STEM.

Typically, STEM education is reflected in one of two ways. Some public school districts will develop a STEM strand within an existing high school curriculum. For example, a school may offer Project Lead the Way, which is a series of preengineering courses (Kelley, Brenner, & Pieper, 2010). In oth-

Improving Urban Schools, pages 7–24

er cases, stand-alone STEM schools are created. As implied, STEM schools focus on STEM areas within traditional high school content. Students in STEM schools meet typical state-level graduation requirements while simultaneously having greater options related to STEM courses coupled with greater academic rigor in all courses (Bell, 2010). Legislation and policies regarding STEM-focused high schools are predicated upon the hope that their availability will increase the percentage of high school students who enter the higher education pipeline anticipating a STEM career and exit the pipeline as STEM professionals. While this may inherently occur given the higher need within STEM fields, STEM schools with programs or structures that promote college and career readiness may contribute to student success in careers and higher education endeavors, regardless of students' socioeconomic status, gender, and/or ethnicity.

This chapter will synthesize current literature regarding college readiness and its interpretation by students, parents, and higher education faculty. We will also review student-engagement literature regarding the relationship between engagement, college readiness, and college persistence or graduation. Finally, presented here are summaries of career exploration tools and analyses that empower students to explore the nexus of their skills, interests, and academic achievement for career directions

STEM PHILOSOPHY

One of the foci of STEM education, via a STEM school or a STEM strand within a school, is the inquiry learning process (Bell, 2010) with interdisciplinary content delivery. While working in collaborative or cooperative groups, students develop and implement a plan to address an identified problem, to record and analyze related data, and to share the results with peers and the community (Sormunen, 2008). Typically, these nontrivial problems are generated from student interest and/or community needs, enhancing students' investment in the process. Given that the problems studied are of interest to the community, it is common for students to present their solutions to school and community stakeholders, thus fostering partnerships between the school and area businesses and/or institutes of higher education (IHE). These partnerships are essential, as they allow students the opportunity to access STEM fields in authentic situations and receive critical feedback from STEM professionals. Students incorporate the feedback into their final product. The inquiry process facilitates the practice of *Project-Based Learning* (PBL), which is so common within STEM education (Bell, 2010). In general, PBL promotes students' active engagement in authentic situations (Bell, 2010) via hands-on learning practices (Steffe & Gale, 1995).

College Readiness

The authors use the term *college ready* rather than *college and career ready* for simplicity. College ready describes high school graduates who are prepared to make a contribution to society, either in a career or in college. In fact, the American College Testing program (2005) stated that there is "evidence that college readiness also means workforce readiness" (p. 5).

Historically, college readiness referred to a student's performance on tests of core academic disciplines such as reading, writing, mathematics, science, social studies, and the arts. These include the American College Testing program (ACT, 2011b) and the Standardized Admissions Test (SAT, 2011), both of which are assessments required for entrance to a variety of IHEs. This continues to be a common practice in spite of the fact that as early as 1994, it was recognized that these and similar test scores were not strong predictors of college success (Armstrong, 1999; King, Rasool, & Judge, 1994). More specifically, Armstrong (1999) found "little or no relationship between [college placement] test scores and student performance in [college] class[es]" (p. 36).

Although test scores may classify an applicant as *college eligible*, this does not equate to one being *college ready* (Lombardi, Seburn, & Conley, 2011). A student who graduates from high school and is truly ready for college will be able to succeed in credit-bearing first-year college classes at a level that provides the background and skills necessary for the next course in a sequence or the next level course in the subject area.

While the percentage of U.S. students who are identified as being college ready is rather low according to a number of sources, the statistics are confusing, because they are generated from various self-report surveys. For example, using the National Assessment of Education Progress (NAEP) High School Transcript Survey data, Greene and Forster (2003) documented that less than 33% of students were college ready in 2001. In 2004, the National Center for Education Statistics (NCES) documented that 30% to 60% of entering freshmen were required to complete remedial courses before taking credit-bearing coursework (Wirt et al., 2004). This statistic is based on responses to the NCES Common Core of Data, which includes information from all public elementary and secondary schools, all local education agencies, and all state education agencies throughout the United States. More recently, after examining high school grades and ACT performance, the ACT reported that only 23% of the nation's 2009 graduating seniors (n = 1,480,469) were likely to be successful in entry-level credit-bearing courses at a college or university (ACT, 2009).

Although Greene and Forster (2003), Wirt et al. (2004), and ACT (2009) used varying calculations and data sources between 2001 and 2009, all reported low percentages of high school graduates' college readiness. In general, it appears that only about one third of U.S. high school graduates are ready to be successful in higher education in spite of all of the high school reform efforts.

Theoretical Dimensions of College Readiness

The concept of college readiness is multifaceted. Besides academic performance, there are additional characteristics and behaviors that are identified as being highly correlated with college readiness or postsecondary success. For example, Byrd and McDonald (2005) studied college readiness among first-generation college students in a community college setting; all participants had work experience prior to entering college. The researchers identified three college-readiness skills needed in addition to content knowledge: time management, goal focus, and self-advocacy skills. Meanwhile, Roderick, Nagaoka, and Coca (2009) identified four essential sets of skills related to college readiness: content knowledge and basic skills, core academic skills, noncognitive or behavior skills, and college knowledge. Furthermore, Conley (2005, 2007, 2008, 2010) posited four dimensions of college readiness: cognitive strategies, content knowledge, academic behaviors, and contextual skills and awareness. The authors will use Conley's dimensions as a framework within this chapter.

In Conley's dimensions of college readiness, *cognitive strategies* are those that enable students to understand and apply content knowledge in order to make inferences, interpret results, develop explanations and solutions, and to think critically about the content; metacognition is key for these strategies. *Content knowledge* refers to knowledge about all academic disciplines (reading, writing, mathematics, science, social science, languages, and arts). *Academic behaviors* refer to the skills, habits, and attitudes students need in order to successfully respond to the rigor and workload typical of college coursework; these skills are forms of self-management. Finally, *contextual skills* refer to an awareness of the college culture, including admissions and financial aid; these skills also include understanding how colleges differ in academic specialties such as agriculture or engineering specialties. In summary, college readiness not only refers to content knowledge, but also includes study skills, self-awareness, and college awareness.

Student and Parent Perspectives of College Readiness

Reid and Moore (2008) captured student voices of first-generation minority college freshmen regarding supports and barriers experienced during their transition into higher education. Students' home-life and their high school experiences, in addition to specific college campus components, were identified as sources of support and challenges. Students noted that advanced placement courses, English, and other demanding academic coursework in high school prepared them for the rigor of college classes, that is, *content knowledge.* These courses also facilitated their development of *cognitive strategies* and *academic behaviors* aligned with their learning styles before entering college. One student advised high school students to "Take AP classes. TAKE THEM TAKE THEM TAKE THEM! It will help you so much" (p. 247). Students struggling in their first year of college believed that not having study skills was a major barrier: "It was so easy not to study because we had the worksheets. If I could go back, I'd study more and make my own notes so that when I got to college, I'd be prepared to study already" (p. 254). These skills reflect Conley's *academic behavior* dimension of college readiness. The importance of adult mentors, high expectations from adults, and counseling regarding college choice, application, and financial aid was also noted, all of which reflect the students' *contextual awareness,* as indicated by Conley (2010).

Byrd and McDonald (2005) worked with liberal arts college students regarding their understandings of what it meant to be college ready. The students identified two major components of Conley's (2010) dimensions of college readiness: college readiness skills and abilities, and background. For instance, one participant stated, "When I think of what it takes to be ready for college, I think it's more of a mental mind set... of having a goal" (p. 29), describing Conley's *academic behavior* and *cognitive strategies.* Furthermore, participants shared anecdotal evidence of how they advocated for themselves when interacting with their professors and how such self-advocacy facilitated their academic success. Finally, participants noted the importance of *contextual skills.* For example, one student said, "It's amazing; I mean I could have gone to college right after high school, but I really didn't know that" (Byrd & McDonald, 2005, p. 30), not realizing that loans were available for college.

STEM Schools and College Readiness

The authors of this chapter were part of a research study that included focus groups with high school students in grades 8 through 12 at two Midwestern STEM schools and a focus group with parents of students in the STEM schools. During these sessions, participants described college readiness as multidimensional; in some ways, their comments aligned with Conley's (2010) work (manuscript in preparation). Both students and parents included the need to be academically prepared for the core disciplines, aligning with Conley's dimension of *content knowledge.* Students also described college readiness to include behaviors such as time management, metacognition, persistence, and the ability to collaborate; these behaviors align with Conley's cognitive and noncognitive dimensions. One student succinctly summarized an important aspect of cognitive and noncognitive dimensions as follows:

> Not everything in college is going to go your way, you're not always going to have a teacher that teaches you in the style that you need to be taught in, and you're not always going to have peers that agree with you and that can help, so you need to learn how to fix yourself in order to succeed. (Student focus group, May, 2011)

Another student posited that "You need to be well organized; if you're not, a lot of stuff can get messy, you can get behind. Also, the environment you put yourself in" (Student focus group, May, 2011). These quotes reflected students' acknowledgement that there are behaviors outside of the cognitive dimension that contribute to college readiness.

Parents felt college readiness also included the need for students to be more aware of the outside world, to be cognizant of the support available through colleges, both financially and academically, and to focus on their long-term career goals, all of which align with Conley's (2010) *college knowledge* dimension. Regarding the outside world, one parent suggested that successful students should experience "Opening your mind; here's the world" (Parent focus group, May, 2011). Another parent suggested that being successful in college is "a mindset that I think that they have to become ready to be their own advocate" (Parent focus group, May, 2011). Another stated that students should "know how to navigate the system, how to enroll for classes, how to find where they need to be" (Parent focus group, May, 2011). For parents of high school students, college knowledge is an integral part of readiness. Together, the students and parents described the four general behaviors and skills identified by Conley as dimensions of college readiness.

As described above, STEM schools naturally embody three of Conley's (2010) four dimensions of college readiness: *cognitive strategies* through the inquiry-based PBL instruction; *content knowledge* through interdisciplinary curriculum; and *academic skills* through collaborative and cooperative structure. College *context skills and awareness* are embodied in the typical partnerships with IHEs and businesses as well as through typical high school counseling. An appropriate analogy might be that the STEM school philosophy provides the growing medium or culture for the continued development of students' college readiness.

Student Engagement and College Readiness

Defining Engagement

According to Furrer and Skinner (2002), student engagement reflects "active, goal-directed, flexible, constructive, persistent, focused interactions with the social and physical environments" (p. 149). Students who are engaged take more initiative, are more curious, and put forth greater amounts of effort compared to disengaged students (Skinner & Belmont,

1993). Consequently, engaged students are more likely to attain higher academic achievement (Klem & Connell, 2004), to pursue higher education (Marks, 2000), to have greater self-esteem (Sirin & Rogers-Sirin, 2004), and to display more positive classroom behaviors (Finn & Rock, 1997) than disengaged students. Although engagement has been noted to diminish as students progress through school, it can be promoted by school practices. For example, teacher support (Garcia-Reid, 2007) and students' feelings of being connected with peers (Connell, Halpern-Felsher, Clifford, Crichlow, & Usinger, 1995) have been shown to facilitate student engagement.

While operationalizations of student engagement differ, models such as the High School Survey of Student Engagement (HSSSE) framework typically include cognitive, behavioral, and/or emotional indicators (Skinner & Belmont, 1993; Yazzie-Mintz, 2010). According to the HSSSE framework, cognitive engagement refers to the academic work that students do (*content knowledge*) and how they go about doing this work (*academic behaviors* and *cognitive strategies*). Within this framework, social/behavioral engagement describes the ways by which students interact with the school community, while emotional engagement describes how connected students feel to the school (Yazzie-Mintz, 2010).

While other models reflect relatively slight variations of this framework, connections with Conley's (2010) notions of college readiness can still be noted. Metacognitive strategies and critical thinking (Linnenbrink & Pintrich, 2003), as well as students' self-regulation, persistence, and effort (Miller, Green, Montalvo, Ravindran, & Nichols, 1996) are viewed as being indicative of cognitive engagement. Together, these cognitive indicators represent two of Conley's dimensions of the college readiness. More specifically, the metacognition and critical thinking skills reflect *cognitive strategies*, while the ability to self-regulate and persist represent the college readiness dimension of *academic behavior*; both of these types of cognitive engagement contribute to college readiness. Behavioral engagement is often identified as students' school attendance, in-class participation, and participation in after-school activities (Finn, 1989; Finn & Voelkl, 1993). The behaviors are aligned with Conley's college readiness dimension of *academic behavior*. Meanwhile, emotional engagement occurs when students feel they are a part of the school community and when they recognize the significant role school plays in their lives (Finn & Voelkl, 1993). Students' understanding of the significant role contributes to the *contextual awareness* dimension of college readiness.

Student Engagement in STEM Schools

The engaging hands-on focus and authentic nature of PBL, coupled with the problem solving involved in inquiry, necessitate greater amounts of student engagement. As such, students who experience STEM education might be more cognitively engaged in academic content and problem solving than those who experience more traditional, siloed, high school curricula.

Unfortunately, there is very limited literature on student engagement in STEM education. In 2011, Patel, Franco, and Lindsey (2012) used the HSSSE framework (described earlier) and corresponding student survey to examine student engagement in two open-enrollment STEM schools in the Midwest. Specific focus was placed on the examination of cognitive engagement items within the survey. Overall, results indicated that tenth-grade students were more cognitively and socially engaged compared to students in other grade levels.

Further analyses compared STEM students' cognitive engagement to a national sample of over 42,000 U.S. high school students who completed the same survey. The purpose was to determine if differences existed in cognitive engagement between students in traditional schools and students in open-enrollment STEM schools. Analyses examined differences between STEM students and the national sample across two broad areas of cognitive engagement: (1) the schools' contribution to growth and focus on work and (2) the challenge and effort related to the coursework. Chi squares were calculated across each item related to the broad areas to determine if there were significant differences between students in the national sample and students in the open-enrollment STEM schools. Analyses revealed that all differences related to cognitive engagement were significant at the .01 level. Results revealed that STEM school students were more likely to report that their school facilitated their critical thinking skills, compared to students from the national sample. Furthermore, STEM students were more likely than students from the national sample to report that their school helped them to relate academic content to post–high school work, to speak effectively, and to comprehend material that was challenging. Similarly, STEM school students were also significantly more likely to report that for most, if not all, of their classes, they were required to put forth their maximum effort. This aligned with findings that STEM students and the national sample differed in how important it was to read for class, with the former more likely to note that it was a very important or a top priority. These results highlighted the difference in college readiness between STEM students and those students in the national sample. The greater facilitation of critical thinking by teachers in the STEM schools may reflect the schools' focus on developing students' *cognitive strategies*. Furthermore, the challenging coursework with the STEM schools promoted students' comprehension of more in-depth *content knowledge*.

The differences in cognitive engagement were not surprising, given the focus STEM education placed on PBL and inquiry-based learning, as both require critical thinking about the content (Steffe & Gale, 1995). Additionally, both schools embedded *habits of mind* within their school culture (Costa & Kallick, 2009), which reflected skills that were beyond the scope of direct academic content. The habits of mind typically serve as a schools' foundational culture (Berger, 2006) and may include collaboration, engagement, and critical thinking skills among other skills (Patel et al., 2012). Consequently, these foundations may serve as an impetus for the differences in cognitive engagement between STEM students and those students in the national sample.

Unlike cognitive engagement, there were components of emotional engagement that were more likely to be exhibited by students in the national sample. The Chi Square analysis indicated the difference was significant at the .01 level. Students in the national sample were more likely to feel that "at least one adult in the school knows them well." This was unexpected given the use of advisory classes in both STEM schools; advisories are meant to "deal directly with the affective needs" of students (Beane & Lipka, 1987, p. 40), Meanwhile, STEM students were more likely to feel that most or all of their teachers believed that they could achieve excellent work.

Finally, almost 30% of the STEM and national sample students reported that they would choose another school if they had the opportunity. The chi square analysis indicated that there was no significant difference between the STEM and national sample on this item. These findings were particularly intriguing given the extra steps that families must take in order for students to attend either of the STEM schools in this study. Beyond the application process, many families need to find transportation to send their children to these schools.

With respect to behavioral engagement, the limited data available noted that students in the national sample were much more likely to participate in school-sponsored activities compared to the STEM students. National sample students were more likely to place importance on these activities as well. These findings may be due to the fact that the STEM schools offer limited after-school options. Furthermore, available options typically focus on academics, such as a Science Olympiad team and a robotics team. Although STEM students have the opportunity to participant in activities at their home schools, logistical concerns can make this difficult. These include the start and end times of the schools, distance, and timing of activities, to name just a few.

In summary, STEM school students consistently reported higher cognitive engagement across all of the indicators. They were also more likely to feel support from nonacademic school personnel such as the secretaries and custodial staff. Meanwhile, students from the national sample were more likely to feel that someone in the building knew them and were more likely to engage in after-school programs.

The research focusing on student engagement highlights its importance in the development of college readiness. As noted by Connell et al. (1995), student engagement is the only factor that directly influences students' academic achievement. This achievement represents not only the final outcomes on various assessments, but also the skills, habits, and self-regulatory abilities that necessitate such success. Our understanding of the effectiveness of STEM education can be expanded if we know how engaged students are in STEM schools, particularly with respect to cognitive engagement.

Student Engagement and High School Persistence

According to the U.S. Department of Education Institute of Education Sciences (2010), three specific indicators reflect persistence in education: average graduation rates (those who graduate with a regular diploma); dropout rates; and the educational attainment of individuals between the ages of 25 and 29. The research reveals clear connections between student engagement and academic persistence, particularly with high school completion. The process of disengagement is one that occurs over time rather than representing any particular moment (Stout & Christenson, 2009). Student boredom, lack of academic motivation, and a lack of actively participating in the classroom reflect disengagement from school and related school-life experiences. If such disengagement ultimately facilitates high school dropout, then it stands to reason that high school completion and success is fostered by engagement in school and related school-life experiences (Appleton, Christenson, & Furlong, 2008).

In a study of 443 African American students, Connell et al. (1995) found that for males and females alike, greater student engagement predicted less frequency of school-based risk behaviors. Additionally, males who reported greater school engagement were more likely to stay in school and graduate. Diaz-Booz (2011) noted how emotional engagement facilitated persistence. When examining the new disciplinary plan for one of the poorest high schools in San Diego, she noted that the school used campus beautification tasks rather than suspensions for individuals who did not follow the rules. Additionally, the school took thoughtful and purposeful efforts to use Spanish language communication practices; this was particularly important given that 45% of the families were from a Latino background and 38% were English-language learners. These practices reflected the school's attempt to promote student and families' emotional engagement; such practices make individuals feel more connected to the school and to the school community. As a result of these and other improvements, students' test scores in mathematics and English have been consistently above 90% and 70%, respectively. These are well above the state average.

In order to examine persistence in the two STEM schools studied by Patel et al. (2012), the data across the schools were disaggregated. Results from School A (which has not yet had a graduating class) indicated that most of the students persisted and chose to stay at School A. In its first year, approximately 17% of students left the school at some point during their ninth-grade year. Another 10% of the original ninth-grade class chose not to return to School A for their tenth-grade year. In School B, anywhere from 3% to 10% of the students decided not to return each year since its inception in 2006. The percentage of students that chose not to return each academic year has steadily increased since 2006; in the early years, about 3% to 5% chose not to return, while in the last 2 years, there have been up to 10% of students who have chosen not to return. The higher percentages of students not returning are more typical of ninth- and tenth-grade students; greater persistence was found among junior- and senior-level students. It is important to note that persistence here is focused on staying in a STEM school, not persisting through high school in general. It is highly likely that most, if not all, of these students who left the STEM schools enrolled in schools within their home districts. Furthermore, persistence at the high school level may not operate in the same way at the college level.

Career Exploration

Assumed, but not specifically described as a dimension of college readiness, is student knowledge of personal career interests and skills. Indeed, it is during high school that students begin to differentiate career intents. Career aspirations emerge in the middle school years but are not typically fully formed until late in the high school experiences (Low, Yoon, Roberts, & Rounds, 2005). All high schools have the responsibility of providing counseling and guidance for their students in this regard. Unfortunately, the actual time focused on career-counseling duties is minimal according to a statewide Florida study of elementary, middle, and high school counselors (Osborn & Baggerly, 2004). Education budget challenges that cut or minimize counselor access, along with the ever-growing counselor responsibilities associated with standardized testing and class scheduling, contribute to the challenge of providing students with adequate counselor access.

Beyond time constraints, school counselor knowledge influences a counselor's ability to guide students in career exploration. School counselors are not and cannot be experts about all career opportunities. In fact, most school counselors try to help students recognize their personal skills and interests and then direct students to additional resources for career options related to their general interests (McWhirter, Rasheed, & Crothers, 2000). In a study by McCuen and Greenberg (2009), students reported that school counselors' lack of awareness of emerging STEM-related careers had a negative impact on their choice to pursue postsecondary STEM interests.

Career Exploration in STEM Schools

The inquiry-based and project-based curriculum in a STEM-focused school provides opportunities for students to become familiar with a variety of careers. For example, a possible assignment that reflects this curriculum may involve students working on a project concerning the environmental impact of locations of new businesses within a community. These students would be exposed to the need for professionals such as civil engineers, architects, materials engineers, and electrical engineers. As an inquiry project, students would need access to these engineering professionals and to use engineering tools and methods to collect and interpret information about the placement of new businesses. The project would be interdisciplinary in that students would be required to use mathematics, science, language arts, and technology content knowledge in a real-world project. Moreover, students would work with city planning committees to include the community-specific policies and practices in place regarding locations for new businesses.

The project described above would expose students to a variety of careers. They would not only learn about the careers, but they would also perform specific tasks typical of the profession during the project completion; they would visit sites, access public files regarding property boundaries and restrictions, and determine specifications needed for the new business location. As students finalized their recommendation about the new business location, they would describe their decision processes with a panel of experts. The ensuing discussions would provide students with greater experience related to careers and associated responsibilities. The project-based exposure to careers would be more memorable to students than simply reading about careers. One could say that the STEM inquiry, project-based interdisciplinary work includes stealth career exposure for students. Through connections with IHEs and local business partners, STEM school students have the opportunity to be exposed to higher education careers and careers within their local area. However, for students to maximize their experiences, further exposure to career opportunities is necessary.

Tools for Career Exploration

To assist students in familiarizing themselves with career options related to their interests, numerous career interest and planning tools have evolved. The ACT college entrance assessment offers a career exploration tool for middle school students (EXPLORE) and for high school students (PLAN for tenth graders and the ACT for eleventh and twelfth graders) (ACT, 2011a). An informative career-options student report merges ACT academic achievement results and self-reported interest inventories for each student.

Another resource, KUDER (http://www.kuder.com), is an online survey focused on student career skills and interest that also provides planning services to students. The online tool is designed for students as early as grade 7 and for adults at any point in a career. Student skill and interest data are mapped to either the RIASEC or the 16 Career Clusters.

Franco, Patel, and Lindsey (2012) analyzed STEM high school student responses to KUDER surveys. The student responses were subsequently mapped to the 16 career clusters. Analyses revealed that STEM school students were similar to the average high school student regarding the development of career aspirations over time; the difference was that 44% indicated they had interests in STEM careers as opposed to the national average of 23%.

CONCLUSION

The literature review regarding college readiness indicated that there are four general dimensions of college readiness. These dimensions can be developed and nurtured during the PK–12 experience. For example, instructional strategies that are effective in contributing to student content knowledge should be incorporated in all school experiences. Inquiry and PBL are known to contribute to strong content knowledge. Additionally, academic skills and behaviors should be fostered; college student voices urged high school students to enroll in the challenging classes and to focus on learning how to learn. The same voices also lamented that not knowing about college culture made the transition to college more difficult and sometimes more expensive. A high school experience that includes numerous opportunities for exposure to and understanding of college culture will contribute to high school graduates' college successes.

For many of the college readiness skills, student engagement is a proxy. Only when students are engaged in the learning process will they develop college readiness skills. When working within the context of Conley's (2010) four dimensions, it is evident that cognitive engagement is a key facilitator of college readiness. In order to exhibit *cognitive strategies* developed in learning content, to develop the *content knowledge* fostered by inquiry and PBL, and to use *academic skills* to address the challenge and rigor of collegiate coursework, it is essential that students be cognitively engaged in all associated instructional activities. Consequently, schools seeking to foster college readiness should do so by facilitating their students' cognitive engagement. Schools can monitor student engagement through surveys or anecdotal reports and increase engagement through specific activities when engagement levels drop.

Guided career exploration that allows students to follow their interests and skills is paramount in supporting college readiness. Though tangentially included in exhibited behaviors of student engagement, its importance cannot be emphasized enough. Without career exploration and awareness, students may not be aware of the multitude of careers that are available in a variety of disciplines. They may not develop the insight to consider the development of career fields that do not yet exist. Consequently, they may not fully understand their aptitudes and preferences regarding college and career choices.

RECOMMENDATIONS

By their design, STEM schools more easily embrace and nurture college readiness. Instructional strategies, interdisciplinary curriculum, and stealth career exposure combine to provide rich educational experiences for students. Traditional schools and schools with STEM strands must make concerted efforts to implement the following recommendations to develop engagement and college readiness among graduates. Regardless of school type, the recommendations address national goals of higher graduation rates, increased rigor, and enrollment in higher education.

1. Monitor students' cognitive, behavioral, and emotional engagement, and create opportunities for students to develop all three. Engagement is directly correlated with academic achievement and thus with high school graduation. Schools should offer and promote various after-school activities, including those not related to academics. When students have a voice in the development of activities, they are more likely to be emotionally and socially engaged with the school.
2. Teachers, with administrative support, should create lessons that integrate inquiry and project-based learning in all disciplines. The research around project-based learning documents that students experience increased academic growth when teachers employ PBL. All disciplines can incorporate this instructional method in lesson planning.
3. Embed academic skills and behaviors in all classes for all students, not just STEM courses. Conley's (2010) theory of college readiness isolates the four behaviors and skill categories that are necessary for college success. Schools can merge the four types of behaviors into school-developed habits of mind or learning habits.

4. Nurture partnerships with local businesses and universities to expose students to career and college opportunities after high school. Collaborate with colleges/universities to host campus events that introduce students to the collegiate culture. College knowledge is beneficial regardless of the field in which students are interested. Additionally, support one-on-one relationships between content teachers and area business partners. Again, the partnerships provide career awareness for students pursuing any career, not just STEM careers.
5. Provide career exploration tools for students to explore their personal skills and interests. Include follow-ups with students to discuss the results from the tools.

REFERENCES

American College Testing (ACT). (2005). *Courses count: Preparing students for postsecondary success.* Iowa City, IA: Author.

American College Testing (ACT). (2009). *Measuring college and career readiness: The class of 2009.* Iowa City, IA: Author.

American College Testing (ACT). (2011a). *Homepage.* Retrieved from http://act.org

American College Testing (ACT). (2011b). *About ACT.* Retrieved from http://media.act.org/documents/ACT.History.pdf

Appleton, J. J., Christenson, S. L., & Furlong, M. J. (2008). Student engagement with school: Critical conceptual and methodological issues of the construct. *Psychology in the Schools, 45*(5), 369–386.

Armstrong, W. B. (1999). The relationship between placement testing and curricular content in the community college: Correspondence or misalignment? *Journal of Applied Research in the Community College, 7*(1), 33–38.

Beane, J., & Lipka, R. (1987). *When kids come first: Enhancing self-esteem.* Columbus, OH: National Middle School Association.

Bell, S. (2010). Project-based learning for the 21st century: Skills for the future. *The Clearing House, 83*(2), 39–43. doi:10.1080/00098650903505415

Berger, R. (2006). The fourth and fifth Rs: Respect and responsibility. In T. Lickona (Ed.), *Center for the 4th and 5th Rs.* Retrieved from http://www.educaciondelcaracter.org/files/012408/Ron%20Berger%20article%202007.pdf

Byrd, K. L., & McDonald, G. (2005). Defining college readiness from the inside out: First-generation college student perspectives. *Community College Review, 33*(1), 22–37. doi:10.1177/009155210503300102

Conley, D. T. (2005). *College knowledge: What it really takes for students to succeed and what we can do to get them ready.* San Francisco, CA: Jossey-Bass.

Conley, D. T. (2007). *Redefining college readiness.* Eugene, OR: Educational Policy Improvement Center.

Conley, D. T. (2008). Rethinking college readiness. *New Directions for Higher Education, 144,* 3–13.

Conley, D. T. (2010). *College and career ready: Helping all students succeed beyond high school.* San Francisco, CA: Jossey-Bass.

Connell, J. P., Halpern-Felsher, B. L., Clifford, E., Crichlow, W., & Usinger, P. (1995). Hanging in there: Behavioral, psychological, and contextual factors affecting whether African American adolescents stay in high school. *Journal of Adolescent Research, 10*(1), 41–63. doi:10.1177/0743554895101004

Costa, L., & Kallick, B. (Eds.). (2009). *Habits of mind across the curriculum: Practical and creative strategies for teachers.* Alexandria, VA: Association for Supervision and Curriculum Development.

Darling-Hammond, L., Newton, X., & Wei, R. C. (2010). Evaluating teacher education outcomes: A study of the Stanford Teacher Education Programme. *Journal of Education for Teaching, 36*(4), 369–388. doi:10.1080/02607476.2010.513844

Diaz-Booz, A. (2011). Fast tracking the underserved: A high school succeeds at retaining its large Latino base. *School Administrator, 68*(5), 18–21.

Finn, J. D. (1989). Withdrawing from school. *Review of Educational Research, 59*(2), 117–142. doi:10.2307/1170412

Finn, J. D., & Rock, D. A. (1997). Academic success among students at risk for school failure. *Journal of Applied Psychology, 82*(2), 221–234. doi:10.1037/0021-9010.82.2.221

Finn, J. D., & Voelkl, K. E. (1993). School characteristics related to student engagement. *The Journal of Negro Education, 62*(3), 249–268. doi:10.2307/2295464

Franco, M. S., Patel, N. H., & Lindsey, J. (2012). Are STEM high school students entering the STEM pipeline? *National Consortium for Specialized Secondary Schools of Math, Science, and Technology Journal, 17*(1), 14–23.

Furrer C., & Skinner, E. (2002). Sense of relatedness as a factor in children's academic engagement and performance. *Journal of Educational Psychology, 95*(1), 148–162. doi:10.1037//0022-663.95.1.148

Garcia-Reid, P. (2007). Examining social capital as a mechanism for improving school engagement among low income Hispanic girls. *Youth and Society, 39*(2), 164–181. doi:10.1177/0044118x07303263

Greene, J. P., & Forster, G. (2003). *Public high school graduation and college readiness rates in the United States.* Retrieved from http://www3.northern.edu/rc/pages/reading_clinic/highschool_graduation.pdf

Kelley, T. R., Brenner, D. C., & Pieper, J. T. (2010). Two approaches to engineering design: Observations in STEM education. *Journal of STEM Teacher Education, 47*(2), 5–40.

King, B. W., Rasool, J. A., & Judge, J. J. (1994). The relationship between college performance and basic skills assessment using SAT scores, the Nelson Denny Reading test and Degrees of Reading Power. *Research and Teaching in Developmental Education, 11*(1), 5–13.

Klem, A. M., & Connell, J. P. (2004). Relationships matter: Linking teacher support to student engagement and achievement. *Journal of School Health, 74*(7), 262–273.

Kuo, V. (2010). What is next for the American high school? *Peabody Journal Of Education, 85,* 389–401. doi:10.1080/0161956X.2010.491709

Lee, V. E., & Ready, D. D. (2009). U.S. high school curriculum: Three phases of contemporary research and reform. *The Future of Children, 19*(1), 135–156.

Linnenbrink, E. A., & Pintrich, P. R. (2003). The role of self-efficacy beliefs in student engagement and learning in the classroom. *Reading and Writing Quarterly, 19*(2), 119–137. doi:10.1080/10573560308223

Lombardi, A., Seburn, M., & Conley, D. (2011). Development and initial validation of a measure of academic behaviors associated with college and career readiness. *Journal of Career Assessment, 19*(4), 375–391.

Low, K. S. D., Yoon, M., Roberts, B. W., & Rounds, J. (2005). The stability of vocational interests from early adolescence to middle adulthood: A quantitative review of longitudinal studies. *Psychological Bulletin, 131*(5), 713–737.

Marks, H. M. (2000) Student engagement in instructional activity: Patterns in the elementary, middle, and high school years. *American Educational Research Journal, 37*(1), 153–184. doi:10.2307/1163475

McCuen, R. H., & Greenberg, J. (2009). Educating guidance counselors on engineering as a career and academic choice. *Journal of Professional Issues in Engineering Education & Practice, 135*(3), 91–94. doi:10.1061/(ASCE)1052-3928(2009)135:3(91)

McWhirter, E. H., Rasheed, S., & Crothers, M. (2000). The effects of high school career education on social-cognitive variables. *Journal of Counseling Psychology, 47*(3), 330–335.

Miller, R. B., Green, B. A., Montalvo, G. P., Ravindran, B., & Nichols, J. D. (1996). Engagement in academic work: The role of learning goals, future consequences, pleasing others, and perceived ability. *Contemporary Educational Psychology, 21*(4), 388–422. doi:10.1006/ceps.1996.0028

Osborn, D. S., & Baggerly, J. N. (2004). School counselors' perceptions of career counseling and career testing: Preferences, priorities, and predictors. *Journal of Career Development, 31*(1), 45–59.

Patel, N. H., Franco, M. S., & Lindsey, J. (in press). *Examining student engagement in open-enrollment STEM-focused schools.* Manuscript submitted for publication.

Race to the Top Act. (2011). 112 U. S. C § 844. Available at: http://www2.ed.gov/programs/racetothetop-district/2012-executive-summary.pdf

Reid, M. J., & Moore, J. L. (2008). College readiness and academic preparation for postsecondary education: Oral histories of first-generation urban college students. *Urban Education, 43*(2), 240–261.

Roderick, M., Nagaoka, J., & Coca, V. (2009). College readiness: The challenge for urban high schools. *The Future of Children, 19*(1), 185–210. doi:10.1353/foc.0.0024

Skinner, E. A., & Belmont, M. J. (1993). Motivation in the classroom: Reciprocal effects of teacher behavior and student engagement across the school year. *Journal of Educational Psychology, 85*(4), 571–581. doi:10.1037/0022-0663.85.4.571

Sirin, S. R., & Rogers-Sirin, L. (2004). Exploring school engagement of middle-class African American adolescents. *Youth and Society, 35*(3), 323–340. doi:10.1177/0044118x03255006

Sormunen, K. (2008). Fifth-graders' problem solving abilities in open ended inquiry. *Problems of Education in the 21st Century, 3*(2008), 48–55.

Standardized Admissions Test (SAT). (2011). *History of the tests.* Retrieved from http://sat.collegeboard.org/about-tests/history-of-the-tests

Steffe, L. P., & Gale, J. (1995). *Constructivism in education.* Hillsdale, NJ: Erlbaum. doi:10.2307/749582

Stout, K. E., & Christenson, S. L. (2009). Staying on track for high school graduation: Promoting student engagement. *The Prevention Researcher, 16*(3), 17–20.

U.S. Department of Education Institute of Education Sciences. (2010). Retrieved December 21, 2011 from http://nces.ed.gov/programs/coe/2010/supnotes/n06.asp

Wirt, J., Choy, S., Rooney, P., Provansnik, S., Sen, A., & Tobin, R. (2004). *The condition of education 2004* (No. 2004-077). Washington, DC: National Center for Education Statistics.

Yazzie-Mintz, E. (2010). *Charting the path from engagement to achievement: A report on the 2009 high school survey of student engagement.* Bloomington, IN: Center for Evaluation & Education Policy.

CHAPTER 3

THE PRIVATE SECTOR, BUILDING STEM PARTNERSHIPS, AND MOVING MODELS FORWARD

Robert M. Capraro
Mary Margaret Capraro
Texas A&M

Michael Muzheve
Texas A&M-Kingsville

THE PRIVATE SECTOR

The private sector is a term used to describe business enterprises run by private individuals or groups and that are not controlled by the state. They are sometimes referred to as the citizen sector. By contrast, businesses that are part of the state are part of the public sector.

The main types of businesses included in the private sector are

- Sole proprietor
- Limited or unlimited liability partnerships

Improving Urban Schools, pages 25–38

- Private limited company—private shares with limited liability
- Public limited company—shares that are available to the public. Two examples are
  - Franchise—business owner pays a corporation to use their name and receive specs for the business
  - Workers' cooperative—all workers get identical salaries and make shared business decisions

There are several considerations when weighing the importance of the private sector in supporting STEM education. Projections show that 80% of job positions generated in the next decade will require math and science skills. There is currently a gap between the workforce demands and preparation of students to fill these roles (Ritz & Bevins, 2011).

One way in which the private sector can support STEM is by forming collaborative partnerships with schools (LaFrance, 2010; Machi, 2008). Some factors that contribute to successful collaboration between the private sector and public schools are

1. Identifying a contact person: networks succeed when business partners assign a liaison person whose regular job duties focus on networking.
2. Developing an attentive attitude: businesses react best to requests from individuals who are aware of their corporate priorities.
3. Clearly defining a role for businesses to achieve: allowing students to shadow an employee or providing an information booth at a career fair.
4. Structuring an effort to plan meaningful experiences of bringing school and business partners together: developing internships or other such collaborative efforts.

There can be challenges to initiating, developing, and maintaining these collaborative partnerships with businesses:

1. Participation of the business community
2. Tension between planning and spontaneity
3. Sustainability of efforts when funding expires
4. Assessing impact and moving toward scale-up

There are several examples of private sector efforts to support STEM education (Harper-Taylor, 2010). The Carnegie Corporation of New York provides one example of private-sector support in STEM education. This private-sector organization plans to raise $100 million in the next decade to support STEM education. Some 14 organizations led by New York's Carnegie Corporation and Opportunity Equation including Google, the Gates Foundation, and the Dell Foundation recently announced a $22 million funding initiative for STEM teacher support. The portion of this initiative called 100Kin10 intends to recruit and prepare 100,000 effective STEM teachers over a 10-year period (Carnegie Corporation of New York, 2012). STEM efforts in schools across the country can use these funds for improving STEM education for *all* students (Robelen, 2012).

To provide a solution to the STEM employment issues, the private sector needs to support STEM in schools. According to Laymon (2011) "by forming partnerships, complementary assets, expertise, and resources can be leveraged to help build a competitive workforce for the 21st century" and allow the U.S. economy to improve by resolving unemployment and ultimately improving the competitiveness of the United States.

The United States currently ranks fourth in the world in terms of technology-based companies such as Microsoft and Amazon; however, the United States ranks only 46th in participation in STEM companies This is troubling, as nearly 25% of future jobs requiring college degrees will be related to STEM fields (Seattle Times, 2012).

The involvement of the private sector in STEM education is welcome given the limited funding resources for new education initiatives. During the second annual science fair held at the White House, President Obama highlighted a series of new administrative plans and support from private partners to foster STEM education. He stated, "When I took office, I called for an all-hands-on-deck approach to science, technology, engineering, and math to make sure these fields get the respect and attention that they deserve... The private sector has answered that call as well" (Robelen, 2012).

Quotes from Private Sector Business Executives

"At Intel, we have seen what young people can do with math and science when they are inspired and well taught. It is up to all of us to stimulate that interest and provide teachers who can guide and nurture students in these critical subjects," said Craig R. Barrett, Chairman and CEO of Intel Corporation.

"For nearly a half century, Xerox has been enabling young people to pursue educations and careers in science, technology, engineering and math," said Ursula Burns, CEO of Xerox Corporation. "President Obama's initiative is not only the right thing for the young people of our country, but essential to maintaining America's leadership position in the global economy."

"The American economy has always depended on innovation; and in a knowledge-based society, there can be no real innovation without an educational emphasis on science, technology, engineering, and math," said Antonio M. Perez, Chairman and CEO of Eastman Kodak. "This is especially important to create a workforce that can succeed in today's rapidly changing economy. Kodak has a long history of supporting education in these disciplines because we know that innovation is the key differentiator for our company and our country."

"We are grateful the Obama administration is focused on America's need to increase its STEM proficiency, and believe this will have positive outcomes for technology-focused businesses like Time Warner Cable," Glenn Britt (CEO and President) said. "We have a need to fill the jobs pipeline with qualified engineers and technologists to continue our tradition of significant innovation in the telecommunications industry."

Partnership for Public Service

According to a recent report issued by the Partnership for Public Service (PPS, 2012), only 2.3% of U.S. STEM graduates are considering postgraduation employment in the federal government. Using data extracted from the National Association of Colleges and Employers' annual survey of graduating seniors, the PPS found that 36.7% of undergraduate STEM majors were looking for careers in the private sector, while 33.9% were planning to pursue graduate degrees.

The PPS report (2012) found that entry-level salaries for federal workers with undergraduate degrees are often not competitive with those offered by private-sector employers. Entry-level federal positions offer salaries in the $35K–$45K range, which falls well below the $60K beginning salary expectations of a meaningful proportion of STEM graduates. The PPS report stated that STEM occupations pay considerably more than other private-sector jobs.

For-Profit Entities

Among the entities that support STEM education are pharmaceutical and healthcare companies, energy companies, Internet and software companies, telecommunications corporations, corporations that offer banking and financial services, and corporations that manufacture networking equipment. Although they support STEM initiatives, which may come in the form of financial assistance, resources, or manpower or may involve societies or select groups of the workforce, most of the support is often targeted toward students as well as mathematics and science teachers. Examples of students' support include scholarships, internships, hands-on training, dropout prevention, education materials, academic assistance, and career advice. Some corporations provide professional development or offer grants to teachers involved with STEM subjects. The following section summarizes some of the initiatives supported by private entities.

Through the AT&T Labs Fellowship Program (ALFP), AT&T offers 3-year fellowships to outstanding underrepresented minority and women PhD students studying computer science, math, statistics, electrical engineering, operations research, systems engineering, industrial engineering, or related fields. In addition to receiving one-on-one mentorship, fellowship recipients spend their first summer of the program interning as part of a research team at AT&T Labs Research.

GlaxoSmithKline (GSK), one of the world's leading research-based pharmaceutical and healthcare companies, supports efforts, specifically Dropout Prevention Summits, whose purpose is to identify strategies, provide workable solutions, and promote improved high school graduation rates, which are especially low for minority students. Another brainchild of GSK is the Opportunity Scholarship program, which awards scholarships as a way of recognizing individuals who overcome adversities and change their lives by pursuing an education. Science teachers wanting to obtain National Board Certification can receive funding from the company through the National Board for Professional Teaching Standards.

In addition to providing internships to students historically underrepresented in the field of technology through the BOLD Internships program, Google, through the Google Summer of Code program, offers student developers stipends to work on various open-source-software projects writing code. The general public can learn about code writing through the online program called Code University. High-achieving high school sophomores and juniors can explore STEM careers while residing and attending classes at six of the nation's top engineering schools for 3 weeks through the LEAD Summer Engineering Institutes. The Computer Science for High School (CS4HS) initiative is supported by grants that can be used by universities to develop 2–3-day workshops for middle and high school computer science teachers. The goal of this initiative is to promote computer science and computational thinking in the middle and high school curricula.

Examples of For-Profit Entities

The Regional Alliance for Science, Engineering, and Mathematics for Students with Disabilities (RASEM) and its successor, *RASEM,*[2] based at New Mexico State University, is an alliance of colleges and universities, secondary schools, and educational organizations in New Mexico and west Texas. In all, 23 two-year and four-year colleges or universities are part of the alliance.

Their mission is to "increase quality participation in STEM by students with disabilities leading to expanded numbers entering a high tech work force" (McCarthy & Misquez, 2003, p. 2). Activities included holding miniworkshops during, which special-education teachers could learn to use and become comfortable with the technology.

In addition to training special-education teachers on how to use technology, funds were also made available, mostly to middle school special education and/or math and science general-education teachers through the Teacher Outreach Projects (TOP) component of the partnership. Teachers were to use the funds to come up with hands-on math and science activities for their students. RASEM also sponsored teacher-development workshops to bring special-education and general-education math and science teachers together to explore methodologies on how to engage students with disabilities when teaching math and science.

Another initiative of RASEM was the creation of a mentorship program in which college students with disabilities mentored middle and high school students. Mentors were required to maintain regular contact with their mentees, and they were brought together through math and science camps. RASEM has forged ties with mathematics and science education specialists through the Texas Rural Systemic Initiative (TRSI), the Southwest Education Development Laboratory (SEDL), the Navajo Nation Rural Systemic Initiative, and the Northern New Mexico Rural Systemic Initiative. One of the goals of establishing these ties is to try and bridge the gap between high school and higher education by providing a platform in which ideas on how to make a difference in math and science pedagogy can be shared (McCarthy & Misquez, 2003).

The Math Science Partnership of Southwest Pennsylvania (MSP) is a partnership that brings together 53 K–12 school districts, four regional education service agencies called Intermediate Units (IUs), and four institutions of higher education. The partnership aims to (a) increase K–12 students knowledge of mathematics and science, (b) increase the quality of the K–16 educator workforce, and (c) create a sustainable coordination of partnerships in the IUs. The Allegheny Intermediate Unit (AIU) is the awardee and within this partnership in which the University of Pittsburgh serves as the project's evaluation team (Steele, Pane, Williams, & Olmsted, 2006).

With the purpose of equipping teachers with pedagogical, content, and leadership skills necessary for them to become effective leaders in their institutions, the project offers professional development to K–12 educators and IHE faculty through academies and seminars. The second intervention strategy involves curriculum alignment as well as pedagogical and course refinement. This is accomplished through the use of curriculum frameworks and by having teachers spend 1 to 2 semesters or a summer on the IHE campuses. The third intervention strategy involves support for and propagation of research-based resources and tools (Williams, Pane, Tananis, Olmsted, & Ciminillo, 2006).

Initiated in 2003, the *System-Wide Change for All Learners and Educators (SCALE)* is a partnership of four urban school districts: the Los Angeles Unified School District, the Madison Metropolitan School District, Denver Public Schools, and the Providence Public School District. The University of Wisconsin's Center for Education Research, California State University Dominguez Hills, and California State University Northridge are the other IHE partners (Scholl, Millar, & Owusu-Yeboa, 2006).

One of the project's goals is to ensure that every K–12 student receives high quality instruction in mathematics and science concepts and also take part in rigorous scientific investigations. Attempts to achieve this goal included professional development to improve teachers' content knowledge (Gordon & Kannapel, 2010).

With regard to IHEs, SCALE's vision is to (a) improve science, technology, engineering, and mathematics (STEM) undergraduate education; (b) improve collaboration between STEM and education faculty regarding preservice programs; (c) improve collaboration between IHE faculty and K–12 districts regarding in-service training; and (d) improve institutional policies and practices that support these activities (Hora & Millar, 2009).

Whose Version of Effective Educational Practice is Being Adopted?

While many STEM businesses find meaningful ways to partner with various educational entities, many businesses take advantage of state, federal, and philanthropic funding intended to support and promote STEM education. Lawyers or business people who have little knowledge of schools and student learning often establish these businesses. Because these groups are not part of established organizations, they often do not have the restrictions or oversight that characterizes higher education and education service centers.

The rules that guide the practices of higher education institutions and educational service centers are charged with providing services to the community and state where they are funded. Therefore, innovations created at these institutions are provided free to the state. For-profit and nonprofit companies established to take advantage of STEM funding are not prevented from taking the materials and building on them and then marketing them to schools. This is an important distinction. Both higher education and educational service centers are owned by the people of the state, and there is no marketing budget allocated as part of the grants they receive to develop the materials and products. Given the ability of for- and nonprofits to market heavily and for high-visibility foundations to garner mass-media attention, one might wonder whether the best practices in education actually make it to the public.

Parental choice and how parents are manipulated in the mass-media blitz is done purposefully and skillfully. Every parent wants their child to have a first class, rigorous, and balanced education. We have never met a parent who asked for a substandard education for their child. In fact, parents are streaming to charter schools based on the promise of that exact thing, a college-prep experience from a public school. So when for- and nonprofits and foundations speak with parents, they may launch several preposterous claims.

Let me discuss the claims. "Public schools do not work, the system is broken, and continuing with what has been done is not going to change the facts." This makes the assumption that public education does not work and has not worked. Remember, this is referring to the educational system that produced, arguably, two of the greatest and most innovative minds of our generation. One a true "imagineer," who changed how we interact with computers, cell phones, and e-readers, while the other changed the way the world by developing the software that powered most of the world's computers for at least 2 decades. Both Steve Jobs and Bill Gates started college, but neither completed it, so their accomplishments were not related to higher education, but were rather an artifact of their K–12 experiences. The educational system does not operate by nor is it a monolithic tanker reluctant at responding to rudder changes. The education system responds to changes and adapts quickly, perhaps the rhetoric from the for- and nonprofits and foundations should be that education changes all too often; but this would be a financial *faux pas,* because it is this change that is providing opportunities for these groups to exert greater influence. Parents need to be aware that what makes it to the public media is there because someone who has acclaim (most often not related to education) is behind it. For example, I do not expect that any parents will read this text. My example here is directed to former MS DOS users. Is it believable that any one person could have stood up and said computer software is broken and we need a new graphical interface, and that it would have gotten press, or that Mr. Gates would have listened? The reality is that most consumers were not technically savvy and deferred to experts. Only a few knew any better and were in a position to ask for a better interface. The personal computer revolution transpired through innovation that occurred secretly. Steve Jobs and Steve Wozniak transformed the computer with a graphical user interface and a mouse, purchased from a company ill-equipped to use it. The innovation was neither consumer demanded nor was it a product of private or public initiative. Innovation occurred when a potential met an opportunity. Apple has continued to innovate ever since. Now, it is most important to remember that even though Apple brought the mouse, point-and-click, or graphical user interface to the market, as well as desktop publishing, the company almost went extinct in the 1990s. Just knowing what to do and executing it well was not enough. Unfortunately, Microsoft had a lock on the market with contracts that required new computers produced by almost all manufacturers to install their software, so changing to another operating system was a costly change for consumers. Eventually, only when consumers demanded portability and a small niche market in portable media arose did Apple again move to corner the market. It is implausible that either of these companies, Microsoft or Apple, would have achieved what they have if outside forces hadn't come to bear. In fact, huge public pressure tipped Apple's hand with their early foray into the light portable computer market with the Newton. These products, or should I say failures, fostered negative stock market news and gossip on Internet sites and in the media that Apple was doomed.

Why do for-profit and some well-meaning not-for-profit groups use philanthropic and state and federal funding to intervene in the educational system? The intervention into the public school system often fails to engage the best and the brightest in the field in favor of outside experts. Imagine what might have happened to Apple or Microsoft if a group of financial experts, lawyers, or doctors received funds from state and federal governments and philanthropists to fix the personal computer in the early 1980s?

Sadly, dollars are spent, which have little impact on the single most important predictor of student success: socioeconomic status, which has been found to be the most important factor in nearly all the studies in which it was considered (Sirin, 2005). Substantial funds are spent on developing new curricula, repackaging old curricula, and building new charter schools. Studies are developed to show the impact of philanthropy, counting innocuous variables like the number of teachers impacted then by the induction of students impacted by the teacher, reduction in absenteeism (not counting in-school and out-of-school suspensions), attendance at before- or after-school tutoring (when it is required), and improved grades (when school policy is changed so no child can receive a grade below 50% on any test or assignment, even when they either do not do the assignment or miss every single question. It is much more difficult to improve a grade when you get a zero on an assignment when you do not do it as compared to getting a 50% for the same thing. The metrics being used simply change public focus, provide smoke and mirrors, and are propagated by a finely honed public relations machine.

Education Groups (Universities, Education Service Centers)

Various other groups have also formed partnerships to improve education. These partnerships often have a university or education service center at its core, with partnerships built around specific expertise and research-based practices (Johnson & Brown, 2009). Unfortunately, these partnerships are rarely funded at the same level as other groups, and the longevity is often short because it is nearly impossible to derive the same income from products and services. With a university and education service center, the products and services are funded through public dollars and often end up in the hands of other entities who use a corporate model to disseminate them. Often, universities and education service centers establish partnerships with businesses, which provides mutual benefits to both entities. In this regard, corporate partners are provided with greater access to schools, while universities and education service centers are provided with broader credibility. These relationships are often long-lasting and extend beyond the initial project.

What Makes a Good Partnership?

The success of a partnership depends on multiple factors. One of the primary factors is recognizing and utilizing the expertise of all parties involved in the partnership (Moreno, 2005; Tomanek, 2005). Other factors that contribute to successful partnerships are (a) clear communication, (b) shared decision making, (c) shared resources, (d) expertise and credibility, (e) flexibility, and (f) recognition of other partners' priorities. Sufficient time is also needed to develop and maintain relationships (Bosma et al., 2010; Bracken, 2007; Cobb & Rubin, 2006). There are specific considerations for partnerships between university faculty and K–12 teachers. To this end, Tomanek (2005) cautioned that partnerships between university faculty and K–12 teachers should not be seen as an instructional relationship in which there is a one-way flow of information, which is something that can easily transpire because one side is viewing this as an expert-novice relationship. Instead, K–12 teachers should be viewed as also having expertise and skills that they may contribute for the benefit of everyone involved in the partnership. Partnerships should also be based on the specific classroom and curricula needs of teachers (Tomanek, 2005). Other important factors include providing teacher professional development at the appropriate level and delivering whatever is promised (Moreno, 2005).

What Does It Take to Maintain Partnerships?

Although funding organizations like the National Science Foundation (NSF) envision collaborations between university faculty, public and private groups, and K–12 schools as lasting beyond the initial funding period, such collaborations are rarely sustainable. The inability for collaboration to provide mutual benefits to partners often results in STEM faculty realizing they receive little benefit from collaboration in terms of their own disciplinary teaching and research (Foster et al., 2010; Yin, 2009). Businesses may also have difficulty justifying the outlay of resources and employee time to shareholders, and philanthropic organizations are often off chasing the next opportunity for media attention.

Although there are multiple barriers to successful partnerships, there are three important factors that promote long-term sustainable partnerships. These factors are as follows: (a) building relationship among the top leaders within each respective organization (i.e., superintendents, vice presidents, chief operating officers, and program officers); (b) persistence and retention of top leaders (changes in leadership often results in a shift in the vision and priorities of the organization); and (c) identifying partners who share similar visions, expectations, and outcomes. This final factor, consisting of shared vision, outcomes, and expectations, is considered together under the category of *common vision*, because the components are too interrelated to parse out.

Having a common vision among all partners builds shared ownership, which is critical to sustaining partnerships. Shared ownership develops through long-term planning, conversations, and interpersonal interactions. However, these relationships are the most difficult to build. It is almost impossible to get to top executives, corporate or educational, without some informal link. Making connections with these key individuals can be challenging without the benefit of kismet or being in the right place at the right time. We find it beneficial to participate in activities that attract both business and educational partners. However, just meeting the right person is not sufficient. It is important to open conversation and learn about the mission and vision that an individual holds for his or her organization. In particular, open dialogue about challenges and solutions should be encouraged. Open dialogue provides benefits to both parties. Finally, engaging in discourse about the use of any resultant data and how each partner can access it and use it is essential (Blank, Jacobson, & Melaville, 2012). Partnerships take a great deal of time to build and attention to detail to maintain.

RECOMMENDATIONS

Establishing authentic and meaningful bidirectional partnerships depends on multiple factors. However the primary factor is developing an understanding of the needs and desires of potential partners. The following is a list of recommendations for developing collaborative partnerships:

1. Build relationships before there is a need.
2. Determine what a business, K–12 school, workforce commission, postsecondary institution, or local community agency might like from you.
3. Understand what outcomes potential partners expect from the partnership. Talk about what happens if those expectations are not met or if they are exceeded.
4. Remember that what some potential partners really want is to be invited in, offered a seat at the table, and be involved in the discussion as a member of the advisory board or community involvement council.

REFERENCES

Blank, M. J., Jacobson, R., & Melaville, A. (2012). Achieving results through community school partnerships: How district and community leaders are building effective, sustainable relationships. *Center for American Progress.* Retrieved from http://www.americanprogress.org/issues/2012/01/pdf/community_schools.pdf

Bosma, L. M., Sieving, R. E., Ericson, A., Russ, P., Cavender, L., & Bonine, M. (2010). Elements for successful collaboration between k-8 school, community agency, and university partners: The Lead Peace Partnership. *Journal of School Health, 80,* 501–507. Retrieved from http://onlinelibrary.wiley.com/doi/10.1111/j.1746-1561.2010.00534.x/full

Bracken, S. J. (2007). The importance of language, context, and communication as components of successful partnership. *New Directions for Community Colleges, 139,* 41–47.

Carnegie Corporation of New York. (2012). *$22m fund to kick-off effort by cross-sector partnership to hire, develop, retain excellent STEM teachers* [press release]. New York, NY: Carnegie Corporation.

Cobb, P. D., & Rubin, B. A. (2006). Contradictory interests, tangled power, and disorganized organization. *Administration & Society, 38,* 79–112.

Foster, K. M., Bergin, K. B., McKenna, A. F., Millard, D. L., Perez, L. C., Prival, J. T., . . . Hamos, J. E. (2010). Partnerships for STEM education. *Science Education, 329,* 906–907.

Gordon, E. M., & Kannapel, P. J. (2010, November). *Sustaining improvement efforts to deepen middle-grades teachers' science content knowledge: The case of system-wide change for all learners and educators, the SCALE MSP in Los Angeles Unified School District.* Retrieved from http://www.mspkmd.net/cases/tck/sustainability/scale.pdf

Harper-Taylor, J. (2010). Siemens Foundation and the STEM challenge. *NCSSSMST Journal, 15,* 10–12.

Hora, M. T., & Millar, S. B. (2009). A final case study of SCALE activities at California State University, Northridge: How institutional context influenced a k–20 STEM education change initiative (WCER Working Paper No. 2009-5). Madison: University of Wisconsin–Madison. *Wisconsin Center for Education Research.* Retrieved from http://www.wcer.wisc.edu/publications/workingPapers/papers.php

Johnson, M. L., & Brown, S. T. (2009). University/k–12 partnerships: A collaborative approach to school reform. In D. B. Martin (Ed.), *Mathematics teaching, learning and liberation in the lives of Black children* (pp. 333–350). New York, NY: Routledge.

LaFrance, G. (2010). Bridging the IT skills gap through industry and academic collaboration. *Employment Relations Today, 36,* 25–30.

Laymon, J. (2011, October 26,). Public/private partnerships the way forward on STEM. *US News Weekly.* Retrieved from http://www.usnews.com/news/blogs/stem-education/2011/10/26/public—private-partnerships-the-way-forward-on-stem

Machi, E. (2008). *Improving U.S. competitiveness with k–12 STEM education and training.* Washington, DC: Heritage Foundation.

McCarthy, W. C., & Misquez, J. E. (2003, November 5–8). *Engineering insights: Developing stem education for students with disabilities leads to k-16 partnerships.* In the proceedings of the 33rd annual Frontiers in Education Conference, Boulder, CO (Vol. 2).

Moreno, N. (2005). Science education partnerships: Being realistic about meeting expectations. *Cell Biology Education, 4,* 30–32.

Partnership for Public Service (PPS). (2012). *Federal leaders face challenges attracting top college graduates to government service* [issue brief]. Washington, DC: Authors.

Ritz, J. M., & Bevins, P. S. (2011). *The connection to the 21st century workforce: Technology and engineering education.* Reston, VA: International Technology and Engineering Educator's Association.

Robelen, E. (2012, February 7). Obama unveils STEM ed. plans at White House science fair. *Education Week.* Retrieved from http://blogs.edweek.org/edweek/curriculum/2012/02/president_obama_hosted_his_sec.html

Scholl, L., Millar, S. B., & Owusu-Yeboa, L. (2006). *Organizational change in an institution of higher education: Improving k–20 math and science education through a university–school partnership.* Madison: University of Wisconsin-Madison. Wisconsin Center for Education Research.

Seattle Times. (2012, March 21). *Strategic private-sector investments in STEM education* [editorial]. Retrieved from http://seattletimes.nwsource.com/html/editorials/2014538719_edit21stem.html

Sirin, S. R. (2005). Socioeconomic status and academic achievement: A meta-analytic review of research. *Review of Educational Research, 75,* 417–453.

Steele, J. L., Pane, J. F., Williams, V. L., & Olmsted, S. S. (2006). *Professional development participation and the role of administrator involvement in the math science partnership of southwest Pennsylvania* (Working Paper No. WR-428-AIU). Pittsburgh, PA: RAND.

Tomanek, D. (2005). Building successful relationships between k–12 and universities. *Cell Biology Education, 4,* 28–29.

Williams, V. L., Pane, J. F., Tananis, C. A., Olmsted, S. S., & Ciminillo, C. (2006). *Math science partnership of southwest Pennsylvania: Year three evaluation report* (Working Paper No. WR-396-AIU). Pittsburgh, PA: RAND.

Yin, R. K. (2009). *Establishing long-term partnerships between k–12 districts and science, technology, engineering, and mathematics (STEM) faculty.* Bethesda, MD: COSMOS Corporation.

CHAPTER 4

CULTURALLY RELEVANT PROJECT-BASED LEARNING FOR STEM EDUCATION

Implications and Examples for Urban Schools

Jamaal Young
Jemimah Young
Christina Hamilton
University of North Texas

The representation of traditionally underrepresented populations (i.e., women, African Americans, and Hispanic students) in science, technology, engineering, and mathematics (STEM) professions is consistently lower than that of their Caucasian and Asian counterparts (Byars-Winston, Estrada, & Howard, 2008). College and university enrollment data suggest that while more students from diverse racial backgrounds pursue undergraduate degrees in STEM fields, African American and Hispanic students disproportionately failed to complete degrees in STEM fields (American Education

Improving Urban Schools, pages 39–65

Council, 2006; Anderson & Kim, 2006; National Science Board, 2007). Specifically, underrepresented students such as African Americans and Hispanics represent 28.5% of the U.S. population and almost 40% of the nation's K–12 enrollment, but only 18% of those receiving STEM undergraduate degrees, 9% of the college-educated workers in science and engineering, and 5% of the STEM doctorate recipients (Hrabowski, 2012, p. 325). Increasing the number of successful STEM professionals from these underrepresented populations is essential to decreasing our overreliance on a foreign-born workforce (Passel & Cohn, 2008). Early awareness is essential to reversing the current enrollment and completion trends for underrepresented students.

High school preparation is consistently found to have a strong association with the success of underrepresented populations of students in the STEM fields (Bonous-Hammarth, 2000; Griffith, 2010; Huang, Taddese, & Walter, 2000; Simpson, 2000). Specifically, the successful completion of Advanced Placement (AP) classes in STEM content areas and high SAT scores are correlated to student resilience in STEM majors (Ost, 2010; Rask, 2010). Furthermore, students who enter STEM majors in colleges typically have an interest in STEM early in their educational journey (Kokkelenberg & Sinha, 2010). To this end, the federal government has devoted substantial time and resources to enrich the precollegiate STEM experiences of tomorrow's workforce.

Multiple investments by states and the federal government have not substantially increased the access and sustainability of STEM education for all students. Successful matriculation to STEM majors and subsequent career paths begins in the classroom. The overwhelming prevalence of low-level tasks contributes to students' lack of understanding of content and poor attitudes toward school (Blumenfeld et al., 1991). Appropriately, if students are not engaged in rigorous and relevant STEM activities, their potential in STEM fields may remain untapped. Creating concrete connection to mathematics, science, and engineering that resonates with the student's background and interest without jeopardizing the integrity of the content may create a pathway to STEM education for all students.

Project-Based Learning (PBL) provides a medium to gain student attention, encourage performance, and secure learning. Culturally relevant STEM PBL can provide a bridge to STEM education for underrepresented populations. The purpose of this discussion is to propose Culturally relevant STEM PBL learning as a bridge to increasing access to STEM education for African American and Hispanic students in urban schools. This chapter provides (a) an overview of PBL in a STEM environment, (b) a framework of culturally relevant STEM PBL, (c) a sample of culturally relevant STEM PBL lessons, and (d) implications for research and practice.

Traditional educational models focus on meticulously addressing a well-defined set of learning objectives in a teacher-led learning environment, while the PBL model begins with a well-defined problem and leaves the approach or task ill-defined. The PBL approach promotes student-group interactions that utilize teamwork and problem solving with scientific inquiry (Krajcik, Czeniak, & Berger, 1999). This design provides multiple entry points for students with different academic backgrounds, which allows for a multitude of instructional possibilities. Early conceptualizations of PBL aligned more closely with a discovery learning approach to instruction. Contemporary PBL instead derives much of its focus from constructivism, situated learning theory, and course integration (Hmelo-Silver, 2004; Zastavker, Ong, & Page, 2006). In the context of STEM education, PBL allows students to draw from the curriculum horizontally as well as vertically. Thus, within a single project, a teacher can structure the activities to encompass all core curricula as well as electives.

PROJECT-BASED LEARNING IN THE CONTEXT OF STEM EDUCATION

PBL within the realm of STEM education incorporates scientific study, technology design, engineering, technology, and mathematical analysis (Lou, Liu, Shih, Hung, & Tseng, 2010). This approach utilizes "projects" as the mechanism to encourage student motivation and to provide a means for demonstrating and explaining the content (Bell, 2010; Ravitz, 2010). Various models of PBL exist, but some common elements are (a) an introduction, (b) definition of learning task, (c) procedure for investigation, (d) suggested resources, (e) scaffolding mechanisms, (f) collaborations, (g) reflections, and (g) transfer activities (Grant, 2002). The theoretical framework for PBL is supported by experiential and constructivist learning theories (Dewey, 1938; Vygotsky, 1978).

According to Brush and Saye (2000), the most effective learning environments allow learners opportunities to negotiate meaning with their peers and promote personal interest. Much of the theoretical foundation of PBL is derived from the constructivist learning theory, which aligns with this notation. Constructivist learning environments promote socially constructed learning through structured interactions and collaboration around a meaningful task (Jonassen, Howland, Moore, & Marra, 2003). Also, a related goal of constructivism is to nurture intrinsic motivation and self-directed learning in meaningful contexts (Chanlin, 2008). These goals support STEM education by creating learning environments conducive to exploration and discovery in the STEM content areas. Although constructivism is the foundation of PBL, cognitive psychology and situation learning contribute considerably to the framework (Barak & Zadok, 2009). These learning theories undergird the socially constructed nature of PBL, which promotes group interactions and student-centered learning as well as several other benefits of PBL for STEM education.

PBL is as an instructional model that is well suited for STEM education because the six major attributes of PBL are relevant to success in STEM education and STEM professions. Harada, Kirio, and Yamamoto (2008) described the six attributes of PBL as (a) problems that require in-depth investigation form the core, (b) instructors facilitate rather than direct learning, (c) students investigate topics that interest them, (d) students collaborate with their peers, (e) students constantly reflect on their learning, and (f) the project culminates in a substantive product. The first attribute of PBL that supports STEM education is the project, which forms the main pedagogical construct for implementing PBL in the classroom in the form of a single multiweek activity or an evolving multiyear endeavor (Ginestie, 2002).

According to the second attribute of PBL, teachers are coaches. This attribute helps to simulate working in STEM professions, wherein project managers guide engineers and scientists rather than directing their task. In the classroom, teachers scaffold instruction using activities that aid in bridging gaps in content knowledge and skills that can make the task more accessible (Bell, 2010). Thus, students learn problem-solving skills in a situated context that provides multiple entry points for a diverse group of learners. The third attribute of PBL is that the student directs the investigation. Academic rigor is fostered in PBL by allowing students to tackle demanding socially and personally relevant task (Van Ryzin & Newell, 2009). Because students have input into the context of the project, they are more vested in the outcome and thus begin to recognize the connection between STEM content and their interests.

Collaboration is another aspect of PBL that supports STEM education. Within the context of PBL, the expectation is that students investigate problems and develop solutions, which they then refine with their collaborators (Grant & Branch, 2005). By engaging in these tasks, students begin to socially construct knowledge and understanding in a cross-curricular context. Because STEM education involves learning across content areas that are traditionally considered rigorous, the team aspect creates a content support system. In each PBL team, students have the opportunity to interact with peer content specialists who can impart their knowledge to the team as the project progresses. Finally, reflection is recognized as a vital component of the PBL process (Grant, 2002; Lamer & Mergendollar, 2010). STEM professions require a reflective workforce that can prevent loss and damage by actively assessing the accuracy and precision of their work through subconscious reflection. Appropriately, reflection is a built-in aspect of PBL that is well established in the instructional model. The effectiveness of PBL as an instructional model is also well researched.

UNDERREPRESENTED AND UNDERSERVED: PBL AS AN INSTRUCTIONAL MODEL FOR ALL

Research suggests several instructional benefits that theoretically support PBL as an effective instructional model (Blumenfeld et al., 1991; Krajcik, McNeill, & Reiser, 2008; Solomon, 2003). Three basic benefits of PBL are the development of (a) group dynamics, (b) problem solving, and (c) communication skills (Hmelo, 1998). Brown and Campione (1996) suggest that PBL develops higher-level thinking skills like problem solving, planning, and self-monitoring. Problem solving is recognized as a major process standard for mathematics and science proficiency (NCTM, 2000; NRC, 2000) of the major STEM content areas. Others contend that PBL is an effective means of developing students' ability to fluidly transfer conceptual ideas across a multitude of learning situations (Scardamalia & Bereiter, 1991). This conceptualization of knowledge suggests students begin to view content in an interconnected manner instead of in isolation. Thus, if students learn scientific notation in physical science, they can recall and apply the same knowledge in the mathematics classroom. Attitudinal effects are also recognized as a benefit of PBL. Several researchers conclude that PBL promotes positive student attitudes and motivation (Boaler, 1997; Dev, 1997). Despite the potential benefits of PBL as an instructional model, many educators remain apprehensive.

Many teachers are discouraged from PBL due to the time demands of projects, especially in the midst of an atmosphere of pacing guides and high-stakes testing (David, 2008). Yet on basic academic proficiency standardized testing, results reveal that students engaged in PBL activities outscore their peers in teacher-directed classrooms (Geier et al., 2008). Boaler (1999) also concluded that students who received PBL instruction outperformed students who received direct instruction on standardized tests. In Iowa, several schools moved from "well below average" to "well above the district average" after implementing PBL over a 3-year span (Thomas, 2000). Furthermore, research substantiates PBL as a means to improve standardized test performance by increasing understanding of concepts (Boaler, 2002; Mergendollar, Maxwell, & Bellissimo, 2007; Walker & Leary, 2008).

Underrepresented populations of students consistently underperform on academic task and drop out of school at higher rates than other populations (Chapman, Laird, & Raman, 2010). Disengagement and educational apathy are possible explanations for this phenomenon. Many students who drop out report that the curriculum is disconnected from real life, and those classroom environments are impersonal systems where no one cares (Wagner, 2002). PBL may serve as a means to promote classroom engagement and achievement of underrepresented populations of students. Reversing student attitudes toward school can support student retention and achievement.

The PBL instructional model surfaced within the last several decades as a means to meet the diverse needs of today's learners (Katz & Chard, 2000). However, underrepresented populations of students in high-need urban areas are traditionally exposed to traditional direct teaching methods. This type of instruction is characterized primarily by didactic, teacher-directed experiences that may not necessitate the skills needed to work in open-ended learning environments (Hill & Hannafin, 2001), while students in affluent suburban areas are more likely to receive PBL and other student-centered models. Motivational theories provide a viable and essential lens with which we can examine and address gaps in opportunities to learn (Ruiz, 2011).

PBL is a promising instructional model for underrepresented students for several reasons. First, PBL builds on students' individual strengths and allows them to investigate their interests in the structure of the defined curriculum (Harada et al., 2008). Underrepresented student populations come from diverse backgrounds, which yield diverse and sometimes divergent interests. PBL provides an instructional platform that allows students to approach problems from multiple entry points and perspectives. PBL may serve as a means to contest some of the attitudinal factors that lead to disengagement and underachievement of underrepresented populations of students. One such aspect of PBL is that the model offers rich opportunities to address and support the motivational needs of students (Johari & Bradshaw, 2008). PBL activities address student motivation by contributing to positive increases in attitudinal factors such as academic self-esteem and student dispositions toward education. The theoretical underpinnings of these factors are derived from attribution theory, self-efficacy, and self-determination theory (Johari & Bradshaw, 2008).

The effectiveness of PBL as a mechanism to increase self-esteem and student a disposition is well documented (Boaler, 1997; Fung, 1996; Johari & Bradshaw, 2008; Katz, 1994; Ravitz, 2010). Both self-esteem and student disposition can affect the academic outcomes of underrepresented students. After consistent exposure to PBL activities, students can begin to exude characteristics that suggest a paradigm shift in their view of education. This is characterized by increases in attendance, engagement, and performance on academic task, all of which are consistent with the dispositions of traditional academic high achievment. In a study of students engaged in PBL, Doppelt (2003) concluded that the students' motivation to learn their discipline and willingness to work on their projects for extended periods of time was indicative of the behaviors common to high achievers. PBL is also an effective means of addressing underrepresented student achievement on standardized test.

Several instructional benefits support the development of knowledge and skills that can enhance the achievement of underrepresented populations. PBL activities are especially effective with lower achieving students

(Lynch, Kulpers, Pyke, & Szesze, 2005). Underrepresented students achieve lower on almost all academic tasks; however, PBL may serve as the catalyst to increased performance. Geier et al. (2008) suggest PBL as a viable instructional model to improve student achievement on standardized tests in urban areas. Furthermore, longer retention of content and an increased ability to apply content in multiple situations (Dochy, Segers, Van den Bossche, & Gijbels, 2003) is another benefit of PBL that can support the achievement of underrepresented populations of students. The aforementioned benefits of PBL support the motivational factors and achievement of underrepresented populations of students; however, the purpose of this chapter is to provide a framework for the infusion of culturally relevant PBL as means to support STEM education for underrepresented students. In the following section, a framework for culturally relevant PBL STEM education is presented.

CULTURALLY RELEVANT PBL AS A BRIDGE TO STEM EDUCATION FOR ALL

Culturally relevant pedagogy can serve as a conduit to STEM education for all students. This theoretical model emerged in the mid 1990s as a powerful instructional model for effective education of African American students. Decades later, culturally relevant pedagogy has evolved into an instructional framework to guide the instruction of all students. Ladson-Billings (1992), who coined the term, described culturally relevant pedagogy as a "theoretical framework that attempts to integrate the culture of different racial and ethnic groups into the overall academic framework" (p. 699). Other scholars have contended that culturally relevant pedagogy consists of the schools' organization, policies, and procedures (Ladson-Billings & Tate, 2006; Richards, Brown, & Forde, 2007). Ladson-Billings (1994) further described culturally relevant pedagogy as

> A comprehensive and multidimensional lens of teaching and learning that includes the curriculum, learning context, student-teacher relationships, instructional strategies, and performance assessments that address the whole child. It involves developing students academically and affirming their cultural competencies, while developing their critical, sociopolitical consciousness. (p. 42)

Ladson-Billings and Tate (2006) highlighted three components of culturally relevant pedagogy: academic achievement, cultural competence, and sociopolitical consciousness. Given that culturally relevant pedagogy emphasizes the linkage among home, community, and school, it is realistic to note that teachers and students might not have congruent home, community, and school experiences. The majority (80%) of teachers of minority students are White, middle-class, females (Sleeter, 2008). Researchers reported that teacher demographics have caused a cultural mismatch between the teacher and the learner (Delpit, 1995; Gay, 2000; Irvine, 2003; Nieto, 1999; Spindler & Spindler, 1994).

Irvine (2003) referred to this cultural mismatch as a "lack of cultural synchronization" (p. 24). She stated that when a cultural conflict exists, miscommunication, confrontation, hostility, alienation, low self-esteem, and ultimately, school failure are the results. Irvine encouraged educators to find a connection between the curriculum and the culture so that all students can achieve success in school.

Conversely, through her qualitative research published in her renowned book entitled, *The Dreamkeepers: Successful Teachers for African American Children*, Ladson-Billings (1994) maintained that successful teachers of minority students do not need to have parallel home, community, and school experiences to the students; however, they have to want to invest in their students' academic success by using effective teaching methods. Similarly, Gay (2000) stated that "similar ethnicity between the teacher and student may be potentially beneficial, but it is not a guarantee of pedagogical effectiveness" (p. 205). The teaching methods that Ladson-Billings and Gay referenced are the core of culturally relevant pedagogy (CRP), which is referred to as culturally relevant teaching or culturally responsive teaching.

Gay (2000) defined culturally responsive teaching as "using the cultural knowledge, prior experiences, frames of reference, and performance styles of ethnically diverse students to make learning encounters more relevant to and effective for them" (p. 29). In a culturally responsive classroom, effective teaching and learning occur in a culturally supported, learner-centered context, whereby the strengths that students bring to school are identified, nurtured, and utilized to promote achievement.

Gay (2000) referred to teachers who use culturally relevant pedagogy as culturally relevant teachers. According to Gay, culturally relevant teachers think deeply about what and how they teach. Focusing on the development of long-term academic goals, they use students' cultures and cite real-life and familiar examples to make curriculum relevant. Specifically, culturally relevant teachers "teach to and through the strengths of ethnically diverse students" (Gay, 2000, p. 29). Ladson-Billings (1994) stated that a highly effective culturally relevant teacher (a) has high self-esteem and a high regard for others; (b) sees him- or herself as part of the community, views teaching as giving back to the community, and encourages students to do the same; (c) sees teaching as an art and views him- or herself as an artist; (d) believes that all students can succeed; (e) helps students to make connections among community, national, and global identities; and (f) sees teaching as "digging knowledge out" of students.

According to Nieto (2004), prior to becoming a culturally relevant teacher, a teacher must become a culturally responsible person. That is, teachers should embrace the cultures of their students as well as identify their own cultural biases so that their insecurities will not hinder their ability to give every student an opportunity to learn. Vaughn (2004) stated that teachers' attitudes about cultural diversity are indicators of how cultural responsive they will be in the classroom. Similarly, Banks (2001) stated, "If teachers are to increase learning opportunities for all students, they must become knowledgeable about the cultural backgrounds of their students" (p. 6).

Further, when teachers become aware of their own cultural bias and cultural identity, they become advocates for individuals who are from nonmajority cultures (Bennett, 2007; Guyton & Wesche, 2005). These teachers reportedly demonstrate high levels of personal efficacy because they believe that they can make a difference in the academic and social lives of their students. Moreover, they are cognizant of their own cultural identities and the cultural identities of others (Guyton & Wesche, 2005; Tucker et al., 2005).

The description of CRP varies across studies as researchers elicit their individual interpretations. Young (2010) suggested that many researchers tend to focus on culture in their descriptions of CRP. For instance, CRP is described as a mechanism to strengthen achievement, a tool to validate student life experiences, and a bridge to connect a student's experiences outside of school to classroom experiences (Boutte & Hill, 2006; Boyle-Baise, 2005; Neuman, 1999). Although these descriptions are valid interpretations of c CRP, for the purpose of this discussion, the theoretical underpinnings of CRP as described in "Toward a Theory of Culturally Relevant Pedagogy" (Ladson-Billings, 1995) are applied.

As CRP began to gain support in the research community, it became apparent that the theory needed refinement and further clarification. To this end Ladson-Billings (1995) published "Toward a Theory of Culturally Relevant Pedagogy" to present a set of theoretical principles to undergird CRP. Utilizing the pedagogical practices of eight exceptional teachers of African American students, Ladson-Billings further explicates CRP. Culturally relevant pedagogy is a "theoretical model that not only addresses student achievement but also helps students to accept and affirm their cultural identity while developing critical perspectives that challenge inequities that schools perpetuate" (p. 469). Three propositions emerged from the research: conceptions of self and others, social relations, and conceptions of knowledge. The teacher beliefs from the study are presented in Table 4.1. These teacher perceptions and actions represent a wide range of ideas that should not be operationalized as the some appropriate elements of CRP, but rather represent a continuum of teacher behaviors conducive to culturally relevant pedagogical practices (Ladson-Billings, 1995).

TABLE 4.1 Ladson-Billings Theoretical Underpinnings of Culturally Relevant Pedagogy

Conceptions of Self and Others	Social Relations	Conceptions of Knowledge
1. All students are capable of academic success. 2. Pedagogy is an art that is unpredictable. 3. Community emersion is important. 4. Teaching is the act of pulling knowledge out.	1. Maintaining fluid student–teacher relationships is important. 2. Create connections to all students. 3. Develop a community of learners. 4. Encourage student collaboration and accountability.	1. Knowledge is not static but rather evolving. 2. Knowledge must be viewed through a critical lens. 3. Passion for knowledge and learning should be apparent. 4. Scaffold or build bridges to learning. 5. Assessment should be multifaceted.

The propositions presented in Table 4.1 were infused with the major instructional components of PBL to create a framework for culturally responsive PBL. A theoretical framework for incorporating culturally relevant pedagogical practice into STEM PBL lessons is presented in Table 4.2.

Table 4.2 provides teachers and teacher educators a framework to consider when designing PBL lessons for diverse populations of students. The framework begins with the first step of the project, which is the driving question, consisting of a well-defined outcome but an ill-designed set of tasks to achieve the goal. When designing a driving question for STEM PBL, the framework suggests four qualities from Ladson-Billings (1995) theoretical underpinnings of CRP that teachers should consider.

Under the driving question section, teachers should consider creating a need to know by (a) conceptualizing teaching as the act of pulling knowledge out, (b) adding elements to the questions that give back to the community or address community issues, (c) creating questions that speak to all students, (d) scaffolding learning through supporting activities and experiences, and (e) exuding a passion for learning. When designing a PBL question, teachers should think of activities that require students to struggle to make connections and draw conclusions without the help of the instructor. Student questions should not be answered without group deliberation and independent research. If student questions remain unanswered, then teachers should rephrase or pose a new questions that will continue the process of "pulling knowledge out" little by little. A strong connection to the community is also an element of culturally relevant pedagogy that is worth considering in STEM PBL. Students struggle to understand how science, technology, engineering, and mathematics are related to their com-

TABLE 4.2 Proposed Framework for Incorporating Culturally Relevant Pedagogy Into PBL STEM Education

	Conceptions of Self and Others	Social Relations	Conceptions of Knowledge
Driving Question or Problem: Teachers should create a need to know by:	• Embracing teaching as a notion of pulling knowledge out; • designing questions or problems that give back to the community or address community issues.	• Creating questions that demonstrate a connectedness with all of the students.	• Designing guiding questions and activities that scaffold, or build bridges, to facilitate learning. • "Exuding a passion for knowledge and learning" (Ladson-Billings, 1995).
Problem-Solving Processes: Throughout the PBL, teachers should engage students in inquiry and innovation by:	• Remaining flexible by viewing "pedagogy as art—unpredictable, always in the process of becoming"	• Encouraging students to learn collaboratively and be responsible for one another.	• Constantly remembering that knowledge is not static; it is shared, recycled, and constructed.
Final Product: Teachers should actively give feedback and promote the revision of the final product by:	• Fostering student self-efficacy.	• Maintaining fluid student–teacher relationships; • developing a community of learners.	• Creating formative and summative assessments that are multifaceted. • "Remind students that knowledge must be viewed critically" (Ladson-Billings, 1995)

munity and why these things matter to others in their community. Thus, adding an element of community awareness to the project provides an avenue for parents and community involvement in the project that creates concrete connections for students to understand how STEM content is relevant to their lives.

If a teacher wants to create a universal need to know, then the driving question should include an element that speaks to all students. Accomplishing a connection to all students is simple, however teachers must first develop a strong relationship with all students. Once the relationships are developed, then teachers can incorporate student interest accordingly. Examples of reaching all students include (a) creating opportunities to present the product or presentation in a variety of ways, (b) addressing multiple learning styles, (c) drawing from pop culture (hip hop culture, skater culture, etc.), and (d) other student interest that the teacher discovers. Culturally relevant pedagogical practices coincide directly with the problem-solving processes and activities associated with STEM PBL.

Throughout the problem-solving process of STEM PBL, some important culturally relevant pedagogical practices include remaining flexible; promoting collaboration and team responsibility; and assisting students in the sharing, recycling, and constructing of knowledge. In an effort not to be remiss, it is important to acknowledge that these practices, as well as the other elements suggested by Ladson-Billings (1994), are more or less universally considered good teaching practices that all exceptional teachers should employ. Thus, it is not surprising that teacher flexibility is a natural component of good PBL practice. Because teachers are facilitators or coaches, they must remain flexible in order to allow for deviations in the design of the project and activities. Teams are also an important element of PBL and culturally relevant pedagogy. Ladson-Billings adds that students should work together toward a common goal and hold one another accountable throughout the process. Although team accountability is important in PBL, it is a major emphasis in the theoretical underpinnings of culturally relevant pedagogy. PBL is an instructional model under the theoretical umbrella of constructivism (Jonassen et al., 2003). Similarly, CRP considers knowledge as constructed through collaboration and reconceptualization of information. In the feedback and refinement processes of STEM PBL, several elements of CRP are suggested.

When incorporating CRP in the feedback and refinement process of STEM PBL, teachers should (a) attempt to foster self-efficacy, (b) maintain fluid student-teacher relationships, (c) develop a community of learners, (d) create multifaceted assessments, and (e) remind students to think critically about solutions. Teachers provide the feedback that is crucial to the quality of the final product or presentation of the PBL. Building culturally relevant pedagogical practices into this process can enhance the performance of all students if brokered with care. Self-efficacy is described by Bandura (1997) as "the belief in one's capabilities to recognize and execute courses of action to produce given attainment" (p. 3). Ladson-Billings (1994) suggested that culturally relevant teachers believe that all students are capable of being successful. This suggests that teachers should expand their conception of culturally relevant STEM PBL by providing feedback and affirmation that fosters self-efficacy. Fluid student-teacher relationships are also important for feedback and refinement. Proper investments in terms of praise and affirmation are necessary in order to facilitate the proper response to critical yet constructive feedback. If student-teacher relationships are not fluid, then teacher feedback could be misinterpreted if not valued altogether. The culmination of the project should promote a sense of communal achievement in light of the hours of refinement and collaboration that preceded the final product. Teachers are the facilitators of these interactions by contributing to the number of iterations and modifications to the project, which bring the teams closer together. As with any instructional model, assessment is vital, thus, in culturally relevant PBL STEM education, it is suggested that teachers create assessments that are multifaceted. The nature of PBL promotes multifaceted learning experiences, thus applying this culturally relevant pedagogical practice is appropriate. Quantitative as well as qualitative assessments are necessary to incorporate multifaceted assessments. Teachers should consider performance assessments, anecdotal notes, interviews, rubrics, and formal exams. Accordingly, these should be spread over the course of the project in a formative manner, as well as at the conclusion, to provide a summative assessment. Critical thinking is a very important element of CRP. Teachers should consider engaging students with feedback that is both critical of the project accuracy along with the social ramifications of the project. At this point, it is important to remind teachers and researchers that this framework should not be implemented as a checklist but rather as a holistic overview of the major considerations for the development and implementation of culturally relevant PBL. Three sample lesson plans that utilize culturally relevant PBL in STEM are presented in the section below. These lesson plans were designed through the collaboration of university professors and in-service teachers to ensure theoretical as well as practical compatibility.

CULTURALLY RELEVANT STEM PBL IN PRACTICE

In this section, a culturally relevant PBL STEM lesson plan is presented. The lesson plan engages the learners in the same project, but the rigor is adaptable for multiple elementary grade levels. This lesson combines the ideals of PBL STEM with the theoretical elements of culturally relevant pedagogy proposed by Ladson-Billings (1995). As mentioned in the previous sections, this lesson by no means encompasses or epitomizes the perfect application of culturally relevant PBL in STEM education. Rather, the intent of this section is to provide a practical application of the theoretical framework proposed earlier. The lesson presented here is appropriate for elementary students grades 1 through 5.

RIDIN' HIGH: DUB RIM/WHEEL DESIGN PROJECT GRADES 3–6

WELL-DEFINED OUTCOME

The student will design a decorative wheel/rim that can replace the standard wheel/rim on a 2012 Cadillac Escalade as well as a 3 to 5 minute commercial to promote the new wheel/rim (see Figure 4.1). At the conclusion of this project, the student will understand the basic parts of a circle and their relationships. The student will also develop a working knowledge of the qualities of mass and the different states. Finally, the students should understand the importance of taxes and government programs connected to public safety.

Figure 4.1 Wheel/rim.

INTRODUCTION

DUB, the original automotive lifestyle magazine, has recruited your class to design a new set of wheels/rims for the 2012 Cadillac Escalade. The magazine's founders coined the term *DUB* to refer to wheels or rims that are 20 inches or larger in diameter. In this project, the students will research the current cost of wheels/rims for Cadillac Escalades and propose a unit cost for their new wheel/rim based on size, attractiveness, and construction material. The student will need to consider the terrain for which the wheel is most appropriate by critically assessing the quality of the roads in their community. The student will then design their wheel/rim on centimeter grid paper and appropriately label all pertinent dimensions. The students will complete the design by determining the calculations needed to manufacture their wheel/rim. The students will complete the project by creating a 3 to 5 minute commercial to promote their new set of rims/wheels.

OBJECTIVES

Mathematics

1. The student will use place value to read, write, describe, and compare whole numbers by researching prices for Cadillac Escalade wheels/rims
2. The student will use geometric tools to construct circles by using a compass to draw a circle on a centimeter grid to serve as the canvas for the wheel/rim design.
3. The student will use standard measurement tools such as a ruler to label the pertinent linear measurements on the wheel/rim.
4. The student will apply formulas when appropriate to determine the circumference and area of the wheel/rim.
5. The student will use problem-solving and mathematical reasoning to determine the appropriate scale factor for their design
6. The student will use computational tools such as calculators to make/check calculations of the circumference, area, and scaled dimensions.

Science

1. The student will identify the appropriate resources used to determine a unit cost of their wheel/rim.
2. The student will use a diagram to model the actual wheel/tire by creating a design to scale
3. The student will accurately use a ruler to label measurement on the wheel/rim design.
4. The student will understand that matter has properties (color, hardness, odor, sound, taste, etc.) by determining an appropriate material to use to construct the wheel/rim.

Social Studies

1. The student will apply decision-making process to financial decision making by developing a unit cost for his/her wheel/rim.

2. The students understands the concept of relative location by comparing the location of their neighborhood in relation to their teammates/group members on a city map

Materials Needed

- Computer with Internet access
- Calculators
- Disposable/digital camera or equivalent device
- Compasses
- Imovie or equivalent video editing software
- Centimeter grid paper
- Colored pencils
- Ruler/yardstick
- Yarn/string
- Notebook/journal

Student Introduction

DUB, the original automotive lifestyle magazine, has recruited your class to design a new set of wheels/rims for the 2012 Cadillac Escalade. It is your job to design and promote the new wheels with a commercial advertisement. In this project, you will research the current cost of a set of 18-inch, 20-inch, and 22-inch wheels/rims for the Cadillac Escalade and propose a unit cost for their new wheel/rim based on size, attractiveness, and construction material. In the decision on the size of the rim/wheel, you must consider the quality of the roads in your community.

Engagement

Introduce the project and then explore this link with the class: http://bit.ly/IERNZB. The link goes through scenes from a DUB car show. The students should now have a better idea about the project. This is the perfect time to assign the first task. Next, go to your school library and have each group research the price of an 18-inch, 20-inch, and 22-inch wheel/rim for a Cadillac Escalade on the Internet. Once each group finds one of each size wheel, have the groups create a PowerPoint presentation with four slides: (a) group name and member names; (b) picture of 18-inch wheel/rim, material, price; (c) picture of 20-inch wheel/rim, material, and price, and (d) picture of 22-inch wheel/rim, material, price. This activity should take approximately one 45 to 60 minute class period.

Exploration

The next class period should begin with each group presenting their rim/wheel PowerPoint. During each presentation, the remainder of the class should record the presenters data in their project notebook. Allow 2 to 3 minutes per group for presentations. Next, allow each group to collaborate and decide how to use the data to determine a wheel pricing strategy. Within each group, assign a notetaker to write down the group idea for later. (The strategies will vary by grade level) Allow no more than 10 minutes.

Pass out two disposable cameras to each group or suggest that each group borrow a camera from their parents. (Many students may have cell phones, which will work just as well.) For homework, have the students take a picture of one of the wheels/rims on their parent's car, a family member's car, or a family friend's car. Each group member should have one picture to use later, thus the camera may need to be rotated within the group.

Transition the class to the next activity by presenting the following YouTube video: http://bit.ly/Ra5lY. The video is of a GEICO commercial involving a pothole. Ask the students, "Have you ever seen a pothole in your neighborhood or community?" Have the students discuss their experiences with potholes and how they think a pothole could affect a rim/wheel on a car.

Bring several different materials to class; some to represent the different stages of matter. Pose the following questions to the class: "Which of these materials would you consider using to make your wheel/rim? Why?" Guide the students in a discussion of the different states of matter and the properties of each state. Extend the discussion by describing the processes that can change the state of matter. Then ask the class, "Do you think that it is possible for the materials used in a wheel and a tire to represent all the states of matter?" Allow the students to discuss the question. For homework, ask each group to find as many potholes in their neighborhood as possible. The students should take a picture of each pothole. These activities can take between two to three 45-minute class periods.

Explanation

Open class with a slide show of all the pictures of wheels/rims the students brought from home. In a small basket, provide each group with a ruler, string/yarn, and scissors. In groups, ask the students to discuss how to measure the perimeter of the wheel/rim in their picture with the given materials. This is a good time to remind students about what perimeter is and how to measure it. Allow the groups 5 minutes to deliberate and write down the possible solution. Many of the students will say that you can just measure the outside with a ruler; be prepared to address the complications of this strategy. Once the students have written down their strategy, allow them to implement the strategy on the picture they took of their parents' wheel/rim. Discuss the perimeter of a circle; mention that circles have a special type of perimeter that is called the circumference. Allow the student to measure the circumference. The circumferences will vary, but pose the following question to the students: "What would we do if we did not have any string?"

Allow the class to present suggestions. Then ask each group to measure the diameter. (Remind the students to use the same units they used to measure the circumference) Provide a pictorial representation for the students to follow. Once each group has their diameter measurement, instruct them to divide the circumference by the diameter. The calculations of π will vary, but they all should be approximately 3.14. Walk around and survey the results, checking for large discrepancies from 3.14. Then call on several students and write the numbers on the board. Introduce π as the ratio between the circumference and the diameter of a circle. Emphasize that for any circle and corresponding diameter, if the circumference is divided by the diameter, it will always equal π or approximately 3.14. Now present the formula for circumference of a circle: $C = D\pi$, where $\pi = 3.14$.

Pass out an age-appropriate compass to each student along with one larger piece of butcher paper per group. On the overhead/document camera, show the students how to use the compass. Have the students construct four circles with varying diameters. This may be a good opportunity to introduce the radius of a circle (depending on the grade level and student capabilities). If so, provide the appropriate explanations as needed. Walk around and anecdotally assess each group's ability to create the circles you suggest. Use these examples to measure their ability to apply the circumference formula to different circles. For homework, have the students measure and record the actual diameter of the wheel in their picture.

Present the class with a map of their local city. Label all of the local neighborhoods on the map. Divide the class into groups based on their neighborhoods or vicinities to one another. Within each group, have the students find the smallest and largest (minimum and maximum) number of potholes documented (for higher elementary grades, consider doing mean, median, and mode). Once each group has their lowest and highest number of potholes, place the intervals (data) on the corresponding neighborhood on the map. Allow the students time to reflect on the data. Then ask the student's, "Which neighborhoods have the most potholes based on the data?" Allow the students to discuss why they think certain neighborhoods have more potholes than others. Then ask the students, "Who is responsible for maintaining the neighborhood streets?" The students should then write a persuasive paper to the City Council or other city officials of their choice, suggesting neighborhood pothole reduction and citing the potential safety benefits.

The students should now have the actual diameter measurements from the cars in their pictures. Have the students calculate the actual circumference of the wheels individually. Then have the students exchange diameters and repeat each group member's calculation for accuracy. The students can now compare the picture circumference and diameter to the actual circumference and diameter. Discuss how the actual measurements are different from the measurements in the picture. Ask the students, "How can we determine how much bigger the actual wheel/rim is compared to the picture?"

Allow the students to respond with suggestions. Discuss the term *scale factor* with the class, then allow the student to find the scale factor for the diameter and circumference. Ask the students, "Should the scale factor be the same?" These activities can take up to a week to complete.

Extension

The knowledge the groups gain from the activities leading up to this point should guide the students in the completion of the project. Give each group a container with a ruler, grid paper, compass, calculator, and markers/colored pencils. Tell each group to discuss and agree upon an appropriate design size for their wheel/rim, then have the students begin designing the wheel/rim. Each group member should create one design, then the group must determine which design to use for the project. Once the group decides on a design, they must create a diagram that labels all the dimensions of the actual wheel/rim based on their design. The group should then review their notes on sizes, prices, and materials to determine a final price for the wheel/rim. To complete the project, the students must create a 3 to 5 minute commercial that includes all the pertinent data a buyer needs to make an informed decision for buying their wheel/rim.

Evaluation

Teachers should consider creating several different assessment measures to implement at each stage of the project. It is suggested here that teachers create an overall rubric to measure the final product of the project; this is introduced on the day the project is first assigned. Using the elements of the rubric, teachers should provide feedback and reminders to ensure student adherence to the rubric elements. Aside from the rubric assessment, teachers should consider formative assessments such as anecdotal notes, interviews, and other evaluation measures to assess student performance on the activities presented throughout the PBL process.

DISCUSSION

The purpose of this chapter was to provide a framework for the infusion of culturally relevant PBL as a means to support STEM education in urban schools. In this chapter, an overview of PBL in a STEM environment was provided. This overview presented the vast general benefits of PBL as well as several benefits for underrepresented student populations. This section of the chapter was written to substantiate PBL as an effective means of STEM education. Although PBL is not the only instructional method available for STEM implementation, its components align with the nature of the

STEM content areas and the STEM professions quite well as explained in this chapter.

In the latter part of this chapter, a case is made for a framework to support culturally relevant PBL in STEM education. Culturally relevant PBL in STEM education is suggested as a bridge to STEM education for all students. Therefore, a framework for implementation in PBL STEM education is presented. This framework is not designed as an assessment tool and does not reflect all culturally relevant pedagogical practices that coincide with PBL. These are more of an impetus to promote further research and reflection on the theoretical considerations for effective culturally relevant PBL in STEM education.

In the final section of the chapter, a sample culturally relevant PBL STEM lesson plan is presented. This lesson plan merges many of the elements of PBL with some of the culturally relevant propositions presented in the framework for integration of culturally relevant PBL in STEM education. This project incorporates elements from all three categories presented by Ladson-Billings (1994): Conceptions of Self and Others, Social Relations, and Conceptions of Knowledge. Specific attention in this project was placed on community emersion, scaffolding guiding questions, remaining flexible, and creating multifaceted assessments. Other elements from the framework are also present, but these are the most apparent. Although theoretical and practical applications are presented throughout this chapter, many questions and implications remain for researchers and educators.

A framework for incorporating culturally relevant pedagogy into PBL STEM education is provided in this chapter, but this framework must be examined and refined to affect change in the classroom. Research is needed to evaluate the influence of culturally relevant PBL in STEM education. Specifically, studies need to investigate teacher implementation of culturally relevant PBL in STEM education. These studies will yield the results needed to expand and refine the current framework. Further, student achievement and attitudinal data must be collected to support or refute the importance of culturally relevant pedagogical practices in STEM education. As the body of knowledge increases in this area, the framework will evolve accordingly. This chapter also has several implications for classroom teachers.

This chapter provides several practical implications for teachers considering incorporating culturally relevant PBL in STEM education. One concern for educators is that, as of yet, a manual or lesson plan database for culturally relevant PBL in STEM education does not exist. Further, PBL is cumbersome and evokes concerns from many teachers because PBL does not adhere to meticulous guidelines and lacks available teacher's manuals with objective-driven lesson plans (Clark, 2006). Recently, Capraro, Capraro, Morgan, and Scheurich (2010) edited a book, *A Companion to Interdisciplinary STEM Project-Based Learning: For Teachers by Teachers*, which consists of 24

PBL lesson plans written by in-service teachers. This is an excellent resource because the lessons presented in the text adhere to the elements of PBL that are considered most effective. Teachers interested in culturally relevant PBL in STEM education should consider the aforementioned text a great foundational resource and then select elements from the framework for incorporating culturally relevant PBL in STEM education to infuse as needed.

RECOMMENDATIONS

According to Jonassen and Grabowski (1993), students must possess motivation to learn, the ability to learn, an instructional environment that promotes learning, and effective instruction. Culturally relevant pedagogy can provide an instructional environment that promotes learning and contends that all students are capable learners. Further, PBL is considered an effective model of instruction and has consistently shown a strong influence on student motivation. Together, these instructional practices can make STEM education and employment accessible to all students. However, this cannot be realized until culturally relevant PBL in STEM education is researched, evaluated, implemented, and accepted. To this end, it is recommended that all educators of underrepresented populations, but those who serve students in urban areas in particular, begin to apply culturally relevant PBL to their current STEM education efforts.

As noted earlier in this chapter, PBL and culturally relevant pedagogy are instructional models that offer substantial benefits to all learners, but underrepresented populations of students in urban area are substantially underexposed to these activities. Teachers should begin to reverse this trend by offering quality STEM experiences through the use of PBL that is culturally relevant. The framework presented in this chapter is sufficient to begin the design and implementation process of culturally relevant PBL in STEM education. However, with any instructional change, teachers should provide a gradual transition from their current instructional practices to culturally relevant PBL in STEM education.

We suggest that teachers begin with a smaller-scale project that is presented over no more than a 1-week period. The lesson plan components for PBL correspond closely to the steps in the 5-E instructional model: Engage, Explore, Explain, Elaborate, and Evaluate (Jackson & Ash, 2011). Thus, it is our recommendation that teachers new to PBL start with the 5-E lesson plan format as a design template. Aside from providing a close correlation to the PBL instructional model, the 5-E lesson plan format also provides a powerful constructivist tool that utilizes a 5-phase instructional design. Within each of the aforementioned phases, teachers can address content and activities related to the steps outlined for PBL implementation presented in the chapter. After a blueprint of the project structure is in place, teachers can use the framework for incorporating culturally relevant pedagogy into the PBL STEM education presented in this chapter in order to begin to address the needs of underrepresented populations of students. Teachers should reflect on the effectiveness of their culturally relevant PBL STEM instruction and repeat several iterations of the design, implementation, and reflection process until they feel comfortable with culturally relevant PBL STEM activities. Collaboration horizontally across content areas and vertically with university professors in education and the STEM content areas is also recommended.

Collaboration is strongly suggested in order to develop an evolving understanding and implementation of culturally relevant PBL in STEM education. Horizontal collaborations between mathematics, science, technology, social studies, language arts, and elective teachers create room for communication that yields more elaborate projects with greater implications for student learning. As more teachers begin to collaborate in the project-design-process classroom, walls are knocked down and knowledge begins to flow between different content areas. In addition, as the number of instructional stakeholders increases, the odds of richer culturally relevant interactions also increases, with the number of varied experiences in collaboration.

Collaboration is also necessary to sustain mutually beneficial relationships between K–12 schools and higher education. We recommend that schools and universities work together to develop and implement culturally relevant PBL in STEM education. Colleges and universities currently fail to retain equitable amounts of students across ethnic groups and gender, but increasing the early exposure to STEM activities for underrepresented students may lead to higher retention of students in the STEM content areas. Teachers would also benefit from the direct connections to STEM professions that are impractical without direct links to colleges and universities that train future STEM professionals. These collaborations support student learning by increasing the rigor and relevance of culturally relevant PBL in STEM education. The utilization of these natural relationships between all parties vested in the promotion of STEM education for all students can support and enhance the educational experiences of underrepresented students in urban areas.

REFERENCES

American Council on Education. (2006). *Minorities in higher education: Twenty-second annual status report.* New York, NY: Author.

Anderson, E., & Kim, D. (2006). *Increasing the success of minority students in science and technology.* Washington, DC: American Council on Education.

Bandura, A. (1997). *Self-efficacy: The exercise of control.* New York, NY: Freeman.

Banks, J. (2001). *Cultural diversity and educating: Foundations, curriculum and teaching* (4th ed.). Boston, MA: Allyn and Bacon.

Barak, M., & Zadok, Y. (2009). Robotics projects and learning concepts in science, technology, and problem solving. *Journal of Technology and Design Education, 19*(3), 289–307.

Bell, S. (2010). Project-based learning for the 21st century: Skills for the future. *The Clearing House, 83*(2), 39–43.

Bennett, C. I. (2007). *Comprehensive multicultural education* (6th ed.). Boston, MA: Pearson.

Blumenfeld, P. C., Soloway, E., Marx, R. W., Krajcik, J. S., Guzdial, M., & Palincsar, A. (1991). Motivating project-based learning: Sustaining the doing, supporting the learning. *Educational Psychologist, 26*(3/4), 369–398.

Boaler, J. (1997). *Experiencing school mathematics: Teaching styles, sex, and settings.* Buckingham, UK: Open University Press.

Boaler, J. (1999). Mathematics for the moment, or the millennium? *Education Week, 17*(29), 30–34.

Boaler, J. (2002). Learning from teaching: Exploring the relationship between reform curriculum and equity. *Journal for Research in Mathematics Education, 33*(4), 239–258.

Bonous-Hammarth, M. (2000). Pathways to success: Affirming opportunities for science, mathematics, and engineering majors. *Journal of Negro Education, 69*(1/2), 92–111.

Boutte, G. S., & Hill, E. L. (2006). African American communities: Implications for culturally relevant teaching. *New Educator, 2*(4), 311–329.

Boyle-Baise, M. (2005). Preparing community-oriented teachers: Reflections from a multicultural service-learning project. *Journal of Teacher Education, 56*(5), 446–458.

Brown, A. L., & Campione, J. C. (1996). Psychological theory and the design of innovative learning environments: On procedures, principles, and systems. In L. Schauble & R. Glaser (Eds.), *Innovation in learning: New environments for education* (pp. 289–325). Mahwah, NJ: Erlbaum.

Brush, E., & Saye, J. (2000). Implementation and evaluation of a student-centered learning unit: A case study. *Educational Technology Research & Development, 48*(6), 79–100.

Byars-Winston, A., Estrada, Y., & Howard, C. (2008). *Increasing STEM retention for underrepresented students: Factors that matter.* Madison, WI: Center for Education and Work.

Capraro, R. M., Capraro, M. M., Morgan, J., & Scheurich, J. (Eds.). (2010). *A companion to interdisciplinary STEM project-based learning: For teachers by teachers.* Rotterdam, The Netherlands: Sense.

Chanlin, L. (2008). A common constructivist goal is to support intrinsic motivation and self-directed learning in a meaningful context. *Innovations in Education and Teaching International, 45*(1), 55–65.

Chapman, C., Laird, J., & Raman, A. K. (2010). *Trends in high school drop-out and completion rates in the United States:1978–2008.* Washington, DC: National Center of Educational Statistics.

Clark C. (2006). Problem-based learning: How do the outcomes compare with traditional teaching? *British Journal of General Practice, 56*(530), 722–723.

David, J. L. (2008). What research says about project-based learning. *Educational Leadership, 65*(6), 80–82.

Delpit, L. D. (1995). *Other people's children: Cultural conflict in the classroom.* New York, NY: New Press.

Dev, P. C. (1997). Intrinsic motivation and self-directed learning in a meaningful context. *Innovations in Education and Teaching International, 45*(1), 55–65.

Dewey, J. (1938). *Experience and education.* New York, NY: MacMillan.

Dochy, R., Segers, M., Van den Bossche, P., & Gijbels, D. (2003). Effects of problem-based learning: A meta-analysis. *Learning and Instruction, 13*(5), 533–568.

Doppelt, Y. (2003). Implementing and assessment of PBL in a flexible environment. *International Journal of Technology and Design Education, 13*(2), 255–272.

Fung, P. (1996). Issues in project-based distance learning in computer science. *Journal of Distance Education, 6*(2), 57–74.

Gay, G. (2000). *Culturally responsive teaching: Theory research, and practice.* New York, NY: Teachers College Press.

Geier, R., Blumenfeld, P. C., Marx, R. W., Krajcik, J. S., Fishman, B., Soloway, E., & Clay-Chambers, J. (2008). Standardized test outcomes for students engaged in inquiry-based curricula in the context of urban reform. *Journal of Research in Science Teaching, 45*(8), 922–939.

Ginestie, J. (2002). The industrial project method in French industry and French schools. *International Journal of Technology and Design Education, 12*(2), 99–122.

Grant, M. M. (2002). Getting a grip on project-based learning: Theory, cases, and recommendations. *Meridian: A Middle School Computer Technologies Journal, 5*(1), Retrieved from http://www.ncsu.edu/meridian/win2002/514/project-based.pdf

Grant, M. M., & Branch, R. M. (2005). Project-based learning in a middle school: Tracing abilities through the artifacts of learning. *Journal of Research on Technology in Education, 38*(1), 65–98.

Griffith, A. L. (2010). Persistence of women and minorities in STEM field majors: Is it the school that matters? *Economics of Education Review, 29,* 911–921.

Guyton, E. M., & Wesche, M. V. (2005). The multicultural efficacy scale: Development, item selection, and reliability. *Multicultural Perspectives, 7*(4), 21–29.

Harada, V. H., Kirio, C., & Yamamoto, S. (2008, March). Project-based learning: Rigor and relevance in high schools. *Library Media Connection, 26*(6), 14–20.

Hill, J. R., & Hannafin, M. J. (2001). Teaching and learning in digital environments: The resurgence of resource-based learning. *Educational, Technology, Research & Development, 49*(3), 37–52.

Hmelo, C. (1998). Problem-based learning: Effects on the early acquisition of cognitive skill in medicine. *Journal of the Learning Sciences, 7*(2), 173–208.

Hmelo-Silver, C. E. (2004). Problem-based learning: What and how do students learn? *Educational Psychology Review, 16*(3), 235–266.

Hrabowski, F. A. (2012). Broadening participation in the American STEM workforce. *Bioscience, 62*(2), 325–326.

Huang, G., Taddese, N., & Walter, E. (2000). *Entry and persistence of women and minorities in college science and engineering education* (NCES Rep. No. 2000-601). Washington, DC: U.S. Government Printing Office.

Irvine, J. J. (2003). *In search of wholeness: African American teachers and their culturally specific classroom practices.* New York, NY: Palgrave/St. Martin's Press.

Jackson, J. K., & Ash, G. (2011). Science achievement for all: Improving science performance and closing achievement gaps. *Journal of Science Teacher Education, 23*(7), 723–744.

Johari, A., & Bradshaw, A. C. (2008). Project-based learning in an internship program: A qualitative study of related roles and their motivational attributes. *Educational Technology Research Development, 56*(3), 329–359.

Jonassen, D. H., & Grabowski, B. L. (1993*). Handbook of individual differences, learning & instruction.* Hillsdale, NJ: Erlbaum.

Jonassen, D. H., Howland, J., Moore, J., & Marra, R. M. (2003). *Learning to solve problems with technology.* Upper Saddle River, NJ: Merrill Prentice Hall.

Katz, L. (1994). *The project approach.* ERIC digest. Champaign, IL: ERIC Clearinghouse on Elementary and Early Childhood Education.

Katz, L., & Chard, S. C. (2000). The project approach: An overview. In J. L. Roopnarine & J. E. Johnson (Eds.), *Approaches to early childhood education* (pp. 175–190). Columbus, OH: Merrill.

Kokkelenberg, E. C., & Sinha, E. (2010). Who succeeds in STEM studies? An analysis of Binghamton University undergraduate students. *Economics of Education Review, 29*(6), 935–946.

Krajcik, J. S., Czeniak, C., & Berger, C. (1999). *Teaching children science: A project-based approach.* Boston, MA: McGraw-Hill, College.

Krajcik, J. S., McNeill, K. L., & Reiser, B. J. (2008). Learning-goals-driven design model: Developing curriculum materials that align with national standards and incorporate project-based pedagogy. *Science Education, 92*(1), 1–32.

Ladson-Billings, G. (1992). Culturally relevant teaching: The key to making multicultural education work. In C. A. Grant (Ed.), *Research and multicultural education* (pp. 106–121). London, UK: Falmer.

Ladson-Billings, G. (1994) *The dreamkeepers: Successful teaching for African American students.* San Francisco, CA: Jossey-Bass.

Ladson-Billings, G. (1995). Toward a theory of culturally relevant pedagogy. *American Educational Research Journal, 32*(3), 465–491.

Ladson-Billings, G., & Tate, W. (2006). *Education research in the public interest: Social justice, action, and policy.* New York, NY: Teachers College Press.

Lamer, J., & Mergendollar, J. R. (2010, September). 7 essentials for project-based learning. *Educational Leadership,* 34–37.

Lou, S., Liu, Y., Shih, R., & Tseng, K. U. (2010). The senior high schools students' learning behavioral model of STEM in PBL. *International Journal of Design in Technology Education, 21*(2), 161–183.

Lynch, S., Kulpers, J., Pyke, C., & Szesze, M. (2005). Examining the effects of a highly rated science curriculum unit on diverse students: Results from planning grant. *Journal of Research in Science Teaching, 42*(8), 921–946.

Mergendollar, J. R., Maxwell, N., & Bellissimo, Y. (2007). Problem-based instruction: A comparative study of instructional methods and student characteristics. *Interdisciplinary Journal of Problem-Based Learning, 1*(2), 49–69.

National Council of Teachers of Mathematics (NCTM). (2000). *Principles and standards for school mathematics.* Reston, VA: Author.

National Research Council (NRC). (2000). *Inquiry and the national science education standards: A guide for teaching and learning.* Washington, DC: National Academy.

National Science Board (NSB). (2007). *A national action plan for addressing the critical needs of the U.S. science, technology, engineering, and mathematics education system.* Arlington, VA: National Science Foundation.

Neuman, S. B. (1999). Creating continuity in early literacy: Linking home and school with a culturally responsive approach. In L. B. Gambrell, L. M. Morrows, S. B. Neuman, & M. Pressley (Eds.), *Best practices in literacy instruction* (pp. 258–270). New York, NY: Guilford.

Nieto, S. (1999). *The light in their eyes: Creating multicultural learning communities.* New York, NY: Teachers College Press.

Nieto, S. (2004). Black, White, and us: The meaning of Brown v. Board of Education for Latinos. *Multicultural Perspectives, 6*(4), 22–25.

Ost, B. (2010). The role of peers and grades in determining major persistence in the sciences. *Economics of Education Review, 29*(6), 923–934.

Passel, J. S., & Cohn, D. (2008). *U.S. population projections: 2005–2050.* Washington, DC: Pew Hispanic Center.

Rask, K. (2010). Attrition in STEM fields at a liberal arts college: The importance of grades and pre-collegiate preferences. *Economics of Education Review, 29*(6), 892–900.

Ravitz, J. (2010). Beyond changing culture in a small high school: Reform models and changing instruction with project-based learning. *Peabody Journal of Education, 85*(3), 290–312.

Richards, H., Brown, A., & Forde, T. (2007). Addressing diversity in schools: Culturally responsive pedagogy. *Teaching Exceptional Children, 23*(3), 64–68.

Ruiz, E. C. (2011). Motivation of Latina/o students in Algebra 1: Intertwining research and reflections. *School Science and Mathematics, 111*(6), 300–305.

Scardamalia, M., & Bereiter, C. (1991). Higher levels of agency for children in knowledge-building: A challenge for the design of new knowledge media. *The Journal of the Learning Sciences, 1*(1), 37–68.

Simpson, J. C. (2000). Segregated by subject: Racial differences in the factors influencing academic major between European Americans, Asian Americans, and African, Hispanic, and Native Americans. *The Journal of Higher Education, 72*(1), 63–100.

Sleeter, C. (2008). An invitation to support diverse students through teacher education. *Journal of Teacher Education, 59*(3), 212–219.

Solomon, G. (2003). Project-based learning: A primer. *Technology and Learning, 23*(6), 20–30.

Spindler, G. D., & Spindler, L. S. (1994). *Pathways to cultural awareness: Cultural therapy with teachers and students.* Thousand Oaks, CA: Corwin.

Thomas, J. W. (2000). A *review of research on project-based learning.* San Rafael, CA: Autodesk Foundation.

Tucker, C. M., Porter, T., Reinke, W. M., Herman, K. C., Ivery, P. D., Mack, C. E., & Jackson, E. S. (2005). Promoting teacher efficacy for working with culturally diverse students. *Preventing School Failure,* 50(1), 29–34.

Van Ryzin, M., & Newell, R. (2009). *Assessing what really matters in schools: Building hope for the future.* Lanham, MD: Rowman & Littlefield.

Vaughn, W. (2004). Prospective teachers' attitudes and awareness toward culturally responsive teaching and learning. *Teacher Education and Practice, 17*(1), 45–55.

Vygotsky, L. S. (1978). *Mind in society: The development of higher psychological processes.* Cambridge, MA: Harvard University Press.

Wagner, T. (2002). *Making the grade: Reinventing America's schools.* New York, NY: Routledge.

Walker, A., & Leary, H. (2008). A problem based learning meta-analysis: Differences across problem types, implementation types, disciplines and assessment levels. *Interdisciplinary Journal of Problem-based Learning, 3*(1), 12–43.

Young, E. (2010). Challenges to conceptualizing and actualizing culturally relevant pedagogy: How viable is the theory in classroom practice? *Journal of Teacher Education, 61*(3), 248–260.

Zastavker, Y. V., Ong, M., & Page, L. (2006, October). *Women in engineering: Exploring the effects of project-based learning in a first-year undergraduate engineering program.* In proceedings of the ASEE/IEEE Frontiers in Education Conference, San Diego, CA.

CHAPTER 5

DISRUPTING INEQUITY

How Policy Change Can Foster STEM Access for Underserved Students (K–12)

Frances R. Spielhagen

Success in providing appropriate and rigorous mathematics curriculum for students across the diverse school situations in the United States has eluded school leaders for decades. Educational policymakers struggle with the challenge of educating all students in a democracy while maintaining adequate standards of excellence. Nowhere has this been more difficult than in mathematics education. Since mathematics study is crucial for both the individual citizens and the nation as a whole, school leaders in the United States have long struggled with providing appropriate and adequate mathematics instruction for all students across diverse school situations. Algebra remains at the core of the dilemma.

However, successful policy revamping in a large Southern district closed the opportunity gap in mathematics and reversed disproportionality in high school mathematics. In the face of skepticism from the educational community at-large and the highly publicized failure of other *Algebra for All* initiatives, one large school district in Virginia implemented a mathematics policy

Improving Urban Schools, pages 67–83

that effectively changed the composition of its eighth-grade algebra classes and led to success among previously absent or underserved minority students. Various efforts to provide algebra instruction for all students, notably in Chicago and in the state of California, have failed over time because of the precipitous way in which the reforms were instituted. The school district in this study initiated the reform by creating an infrastructure on which the expanded eighth grade algebra program rests.

As a result of the policy reform in this district, the large majority of eighth-grade students are succeeding on a state test that is usually taken by ninth-grade students in other parts of the state. Moreover, most students have the algebra credential in place before high school, and many more are taking additional advanced math courses in high school than under the former policy. The students report greater college attendance. In addition, the opportunity gap has been closed, at least going into high school, and the district continues to provide a strong foundation for further growth. Their efforts can serve as a template for other school leaders in other districts who struggle to bring about realistic change in this tense era of accountability.

THE ROOTS OF THE PROBLEM

In 2008, the National Mathematics Advisory Panel soundly endorsed the study of algebra as a cornerstone of mathematics for all "prepared" students. Although the panel's report was subject to enough criticism among policy researchers to warrant an entire issue of *Educational Researcher* (AERA, 2008), the question of what constitutes sound mathematics policy remains a mandate for school leaders at all levels. Moreover, Boaler's (2006) acknowledgement of the extensive range and variety of students in typical mathematics classes focused the challenge for policymakers on policy initiatives that scaffold student success. How can school leaders structure policy that fosters suitable preparation for the study of algebra and promotes maximum success for all students?

The algebra debate continues to rage in the national arena, in state departments of education, and in local school districts. Should algebra be required of all students, and when should they study it? Algebra has long been considered suitable for "college-prep" students, those who might "need it" in future careers. However, once that designation is made, the door has been thrown open for discriminatory and biased decisions that prevent certain students from studying algebra and expecting others to take the course. This is the origin of the opportunity gaps that exist in American education today.

However, this was not always the case. In the early days of American education, algebra was part of the traditional grammar school curriculum available to all students as they progressed individually through the grades.

Students studied algebra when they got to it in the normal course of the curriculum, and very often students studied algebra before they attended high school. The environment of small schools allowed teachers to differentiate by simply meeting the immediate needs of the individual students. The conventional curriculum was straightforward, and teachers provided instruction in the various subjects, particularly mathematics, when the students were ready for the topics that flowed from the study before. As a result, some students studied algebra in eighth grade.

However, in the mid-19th to early 20th centuries, a large influx of students from Europe and the southern United States poured into urban environments, especially in the North. This population swell changed the delivery of curriculum into a design that policymakers felt would meet the varying educational needs and employment prospects of this vast range of students. The face of education was changing, both literally and figuratively, and policy was derived from decisions about the practicality of algebra study in terms of potential employment. Educators began to reserve algebra for those who seemed capable or deserving of studying mathematics in high school. In other words, who would need to study algebra? Of course, those who might pursue science and engineering certainly would need to have studied algebra. This initial standardization of the mathematics curriculum led to the movement of algebra to the high school. Algebra became the next step in one's education; a step that led to further academic study in mathematics and science and the gateway to educational and employment opportunities beyond high school.

At the same time, the comprehensive high school became an important link between public elementary school and the state-supported college or university. However, this path was not open to all students. Promotion from the grammar school to the high school was solely related to the assumed interests and ability of the students. Quite simply, algebra became the domain of the wealthy. More affluent students were more likely to advance to high school and more likely to attend college. Therefore, they "needed" to study algebra. Less affluent students left the educational pipeline after grammar school and therefore did not "need" to study algebra. Regardless of whether they could or could not understand algebra, students not planning to go to high school were not given the opportunity to study it.

This situation became more acute when standards were imposed on the high school admissions process, through the creation of high school admissions tests. By 1855, high school admissions tests became one of the first systemic attempts to measure academic progress on the vastly diverse and growing population of students, particularly in urban areas. These tests ultimately became a sorting mechanism that laid the foundation for the inequities that would be challenged a century later. Prior to these external admissions tests, teachers in typically one-room schools relied on holistic appraisal rubrics. In the quest for predictability in educational outcomes, educational policymakers began to suspect these holistic measures as inadequate and antiquated. As a result, the first standardized high school admissions tests emerged as not only a means of measurement but also of control.

The impetus for this initial standards-based reform movement was to employ high school admissions tests as a tool for raising standards in public schools (Cremin, 1980). Predictably, the content of the tests changed the curriculum of the grammar schools. Increased numbers of students began to take the admissions tests, but to succeed on the tests, they had to master grammar school subjects, basically the "three Rs"—reading, writing, and arithmetic—but not algebra! Ultimately, a larger pool of potential high school candidates and the introduction of the high school admissions tests removed "advanced" subjects like algebra and Latin from the grammar school curriculum. Those subjects became the domain of the high school. How did that happen? High school admissions tests no longer contained algebra questions. Since students were not required to answer test items related to algebra on the high school admissions tests, it became unnecessary to study algebra before entering high school. In effect, the standards for eighth-grade students had been lowered. Prior to the high school admissions tests, non–college bound students might have studied algebra in the old grammar school curriculum. With the new testing system, your presumed station in life determined your course of study.

Over 30 years ago, Cremin (1980) highlighted the discrimination that resulted from the removal of the study of algebra from the grammar school curriculum in the early 20th century.

> The elimination of advanced subjects below the high school meant lost opportunities for pupils unable to advance. Most high school pupils were from the middle and upper classes; thus, removing advanced subjects from grammar schools restricted chances for the talented but less advantaged youth to taste the higher learning. (p. 154)

Of course, this institutionalized inequity was not a secret, even though it permeated educational convention throughout the 20th century. Unwarranted assumptions regarding a student's educational trajectory became the basis of the Civil Rights Movement and the foundation of critical race theory. Over a decade ago, Moses and Cobb (2000) asserted that the study of algebra is a civil right and more than simply the key to advanced study and understanding of technology, because "in today's world, economic access and full citizenship depend crucially on math and science literacy" (p. 5). As evidence, they cited the disturbing statistics that in 1995, Blacks composed only 2.1% of the PhDs in engineering, 1.8% in computer science, and 0.6% in mathematics. The authors maintain that the organization of the mathematics curriculum in the industrial era has caused the elite status of the study of algebra to be not only a barrier to college entrance but also to full citizenship and economic autonomy.

While making its deliberations about the status of algebra in the curriculum, the National Panel discussed the study of algebra as the foundation for all mathematics and science (Bass, 2006). However, deep within the collective understanding of some American educators resides the opinion that students must have a "rare, innate, ability... the math gene" (Campbell, 1996, p. 4) to learn algebra, an ability possessed only by a select few. In 1990, the National Research Council advised that formerly exclusive mathematics courses must become inclusive. "To achieve an equitable society, we must change the algebra course from a filter that screens out segments of our population to a pump that propels all students toward opportunity" (Taylor, 1990, p. 45).

The College Board weighed in on the side of increased access to algebra as a means to increased equity and excellence, based on research (Pelavin & Kane, 1990) that suggested that minority students who master algebra and geometry and have aspirations to go to college actually enroll in college at the same rate as their nonminority peers with those same experiences. Therefore, the College Board's Equity 2000 strategy advocated that all students complete Algebra 1 by the ninth grade.

The rhetoric surrounding mathematics literacy in the United States has caused local school districts to redefine their mathematics curricula in order to create equitable opportunities for all students. The failure of the ninth-grade *Algebra for All* initiatives in Chicago and the backlash to the California State Board of Education's mandate of eighth-grade *Algebra for All* within 3 years have become fodder for opponents of these reforms. Soon after the publication of the report from the National Panel in March 2008, *Education Week* reported that some of the members of the national panel considered the California eighth-grade algebra initiative ill-advised (Cavanaugh, 2008). The most vocal among them, Tom Loveless (2008) of the Brookings Institution, maintains that "misplaced math students," are unprepared for the study of algebra, more likely to come from low-income families qualifying for federal free or reduced-price lunch, and overwhelmingly Black and Hispanic.

THE SOLUTION

In the face of this ongoing debate, and in the absence of a cohesive national initiative, school leaders in a large school district (N = 60,000 students) in the southeastern United States courageously responded to the varied recommendations from the National Council of Teachers of Mathematics (Fennell, 2008; NCTM, 2000) and the National Mathematics Advisory Panel (2008). Over several years, these leaders collaboratively and successfully revamped their district's mathematics policy, and the district now provides strong algebra instruction for all of its students in the eighth grade. There are lessons to be learned in the steps they took to change the mathematics policy in their district, their bold decision to establish an *eighth-grade Algebra for All* policy, and the outcome of closing the opportunity gap for their diverse population.

How did they bring about such a drastic change in business as usual in their district? Over a period of 8 years, school leaders worked closely with an outside evaluator who conducted annual reviews of their data. The leaders employed the annual reports to shape and refine their mathematics policy. They also conducted internal analysis of available data and compared the results with those of the evaluator. As that evaluator, I conducted both quantitative data analysis and qualitative interviews. Ultimately, the school leaders used these data to change the established mathematics delivery policy and how they structured a comprehensive reform that effectively changed the faces of the students sitting in their eighth-grade algebra classes. In fact, after 8 years of restructuring, over 90% of the students in the district study algebra at least by eighth grade, and 97% of them pass the state algebra examination. Over the course of the reform, the district superintendent left to become state superintendent, but, learning of the emerging success of the initiative, the new superintendent endorsed the restructuring.

Three questions guide understanding of how the school leaders engineered this successful change of school culture and disrupted the inherent inequity in their mathematics policy. First, what exactly was the situation that prompted the reform? Second, what measures did they take to change the game for underserved students? Finally, what lessons can be learned from the success of their efforts?

CHANGING THE GAME BY CHANGING THE PLAYING FIELD

In 2008, the National Mathematics Advisory Panel recommended that "All school districts should ensure that all prepared students should have access to an authentic algebra course—and should prepare more students than at present to enroll in such a course by Grade 8" (p. 23). However, 8 years before this important policy statement, school leaders in this district initiated a reform, targeting the inherent inequities in their established mathematics delivery policy. Located in the southeastern United States, this large district of 60,000 students is similar to other school districts across the nation. The district serves students who live in an area of approximately 450 square miles. Two thirds of the population of 30,000 is White or Caucasian/European, approximately one third is Black or African American, and a small percentage (about 3%) is "other," mixed Asian, Hispanic, or Native American.

Like many districts across the nation, long-standing tracking policies had institutionalized inequity in the paths students followed in studying mathematics. Through a complicated system of test scores, course grades, and teacher nomination, some students studied algebra in eighth grade but others did not. Quantitative evidence readily revealed the disproportionality of access among Black students, especially males (Spielhagen, 2006) and of the long-term effects of eighth-grade algebra (Spielhagen, 2007) under the original protocol for algebra access. Table 5.1 shows the link between socioeconomic status (SES), as indicated by free and reduced-price lunch and access to eighth-grade algebra, prior to the reform. In general, school leaders learned that the lower the SES of the school population, the lower the percentage of students studying algebra in eighth grade.

Examination of the composition of the eighth- and ninth-grade algebra classes revealed disproportionality of enrollment of Black students in eighth-grade algebra in terms of their presence in the general population. Figures 5.1 and 5.2 illustrate the population trends in the two algebra courses.

Further evidence emerged when the leaders examined advanced mathematics course-taking and college enrollment related to having studied algebra in eighth grade. Figures 5.3 and 5.4 illustrate the disparities that resulted from restricting access to studying algebra in eighth grade.

TABLE 5.1 Course Taking Patterns by Schools: Grade 8 Algebra 1 vs. Grade 9 Algebra 1

School No.	Percentage Free/ Reduced-Price Lunch	Grade 8 Algebra 1	Grade 9 Algebra 1	Ratio Grade 8 vs. Grade 9
Total		N = 1,441	N = 2,224	1.54
1	.1%	207	182	.88
2	.2%	173	178	1.03
3	.6%	217	165	.76
4	.9%	143	310	2.17
5	11.8%	92	136	1.48
6	18.4%	123	191	1.55
7	18.9%	139	194	1.40
8	19.4%	108	257	2.38
9	22.4%	122	209	1.71
10	27.3%	36	130	3.61
11	40.0%	66	245	3.73
Mean		120		1.54

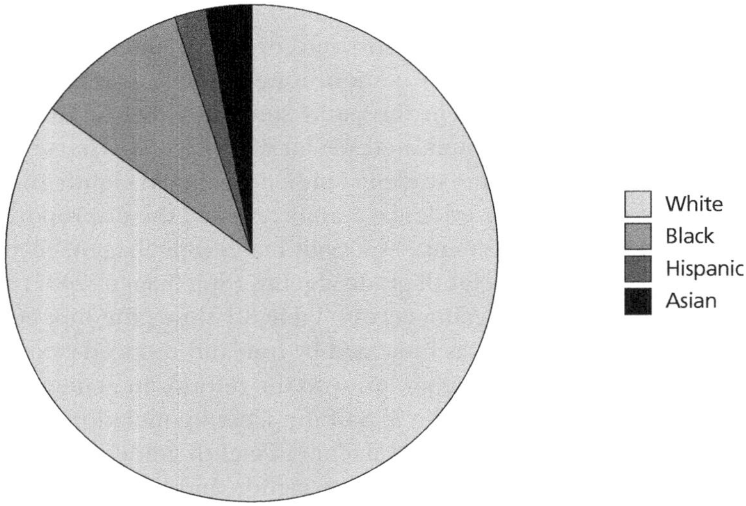

Figure 5.1 Original eighth-grade Algebra 1 enrollment by ethnicity.

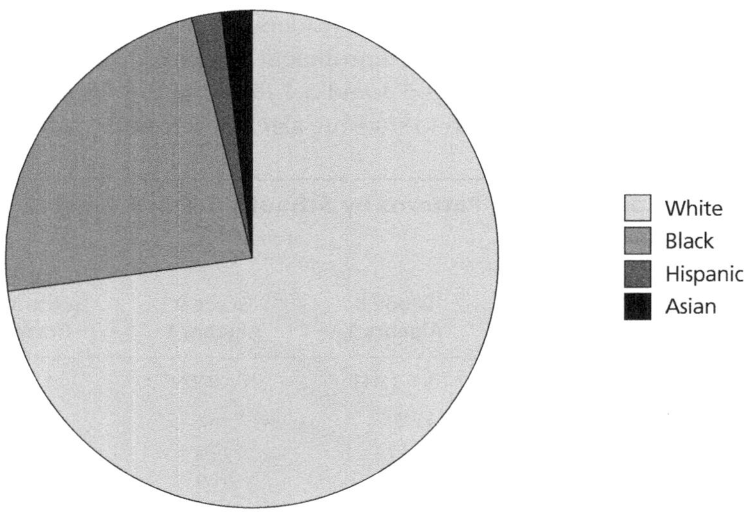

Figure 5.2 Original ninth-grade Algebra 1 enrollment by ethnicity.

Therefore, this evidence prompted the school leaders to change the algebra delivery policy so that after a decade of curriculum organization and professional development and support of the teachers, as well as support for the students, almost all students are studying algebra at least by eighth grade. Gradually but consistently, the schools opened access to eighth-grade algebra to larger numbers of students each year, steadily increasing the number of

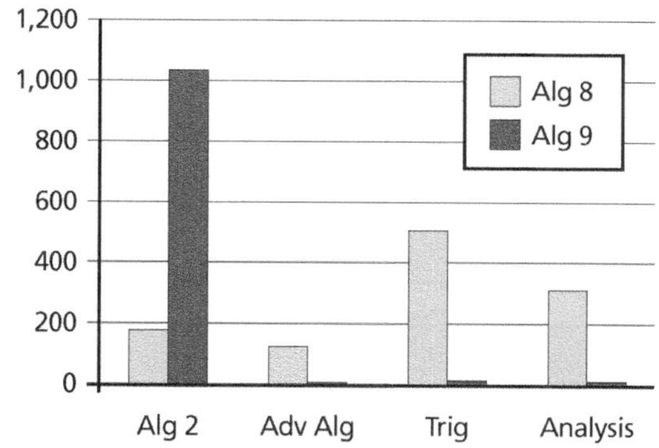

Figure 5.3 Eleventh-grade mathematics course-taking after Algebra 1.

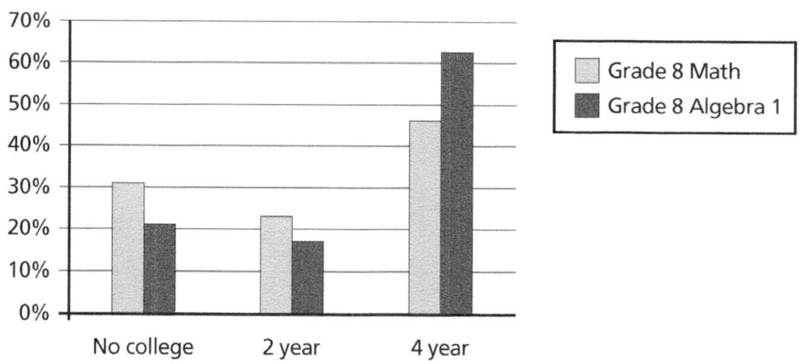

Figure 5.4 College attendance by access to eighth-grade Algebra 1.

students in the eighth-grade algebra classes. Some study it earlier, and some start it in eighth grade and complete the course in ninth grade.

School leaders in this district embarked on a long, diligent process that involved rigorous reflection at all levels, from the classroom teachers to the superintendent. They implemented consistent policy changes gradually and supported all personnel, as well as the students. Student support was critical and involved after-school tutoring sessions as well as differentiation of curriculum to meet the needs of the diverse population. Figure 5.5 illustrates the gradual change in enrollment from 2005 through 2009 in eighth-grade algebra as the district moved steadily toward a successful program of eighth-grade algebra for all but the most disabled students.

After 8 years of working toward this goal, even among those students who cover the algebra curriculum over 2 years (starting in eighth grade and finishing in ninth grade), the vast majority of students in this district are successfully passing the state's Algebra 1 examination (Spielhagen, 2011).

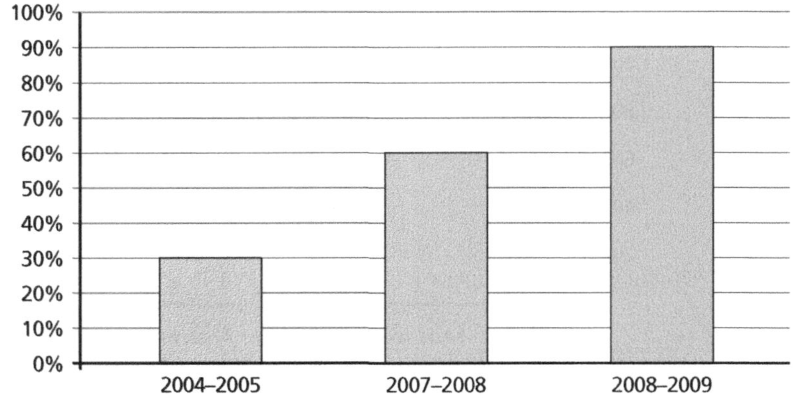

Figure 5.5 Changes in Algebra 1 enrollment, 2005–2009.

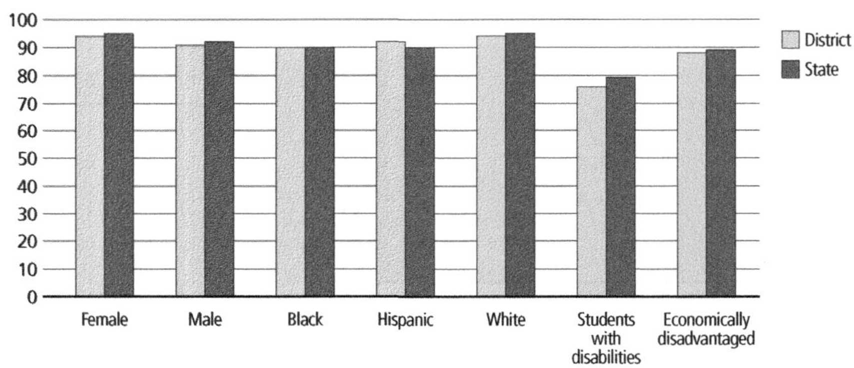

Figure 5.6 Passing rates by demographics, 2009.

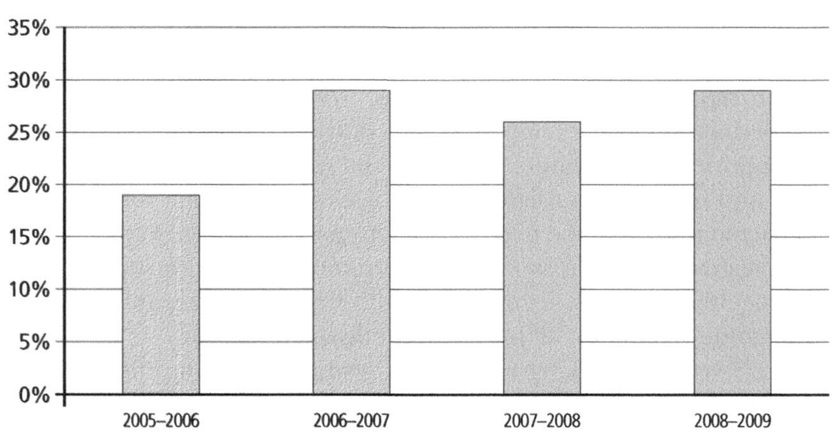

Figure 5.7 Advanced pass rates on state Algebra 1 test, 2005–2009.

Among all eighth-grade algebra students, 92% passed the state algebra test, ranking among other students across the state, students that were in ninth grade! Figure 5.6 shows the pass rates among eighth-grade students in this district in 2009; Figure 5.7 illustrates the changes in "advanced" pass rate on the same Algebra 1 test from 2005 to 2009.

THE COMPONENTS OF SUCCESS

An outspoken and vocal critic of eighth-grade algebra programs, Tom Loveless (2008) suggested four key elements for algebra programs that he has termed "realistic" (pp. 12–13). These elements provide a structure for analyzing the success in this district.

First and most importantly, get the goal right. Focus on learning, not completing a course. Failed *Algebra for All* initiatives, whether they target eighth or ninth grade, fail because the algebra credential is the goal, not the requisite learning. This district established several structures that fostered learning as the goal. Success followed the focus on learning. The mathematics coordinator and school administrators increased time on task in mathematics throughout the intermediate grades. The teachers were shown how to differentiate curriculum to meet the varying needs and styles of their diverse populations. No longer was algebra considered the right and domain of certain groups of students, that is, those expected to attend college. If everyone (or the vast majority of students) was expected to study algebra in eighth grade, then instruction was arranged so as to make it accessible to all students. One teacher proudly explained, "We teach here," disparaging other environments wherein only certain elite students are allowed to study algebra in eighth grade.

Second, Loveless recommended that schools teach and assess prerequisite skills and focus on elementary mathematics leading to algebra. The earliest stages of this reform involved professional development for the third- and fourth-grade teachers in this district. It is no secret that many teachers in the elementary grades across the United States are "generalists," with little special training and background in mathematics. In fact, the common fear is that many of these teachers are also math phobic. By providing targeted professional development in the algebra skills that derive from instruction in the intermediate grades, the school leaders laid the foundation for success for these students in eighth-grade algebra.

Third, engage in early intervention. Build an infrastructure for all students leading to eighth grade. Admittedly, for some teachers in this district, the concept of *readiness* was a difficult stumbling block to the implementation of the eighth-grade *Algebra for All* initiative. At each step of the way, teachers expressed fears about the readiness of their students to handle more complex concepts. However, readiness may be defined in many ways and is a catchall word that can be exclusive rather than inclusive. Gradually, as they continued their own professional development, the teachers realized that they were equating readiness with resources, and that for some students, the teachers may be their best, if not only, resource. As a result, after-school and lunchtime tutoring programs began to emerge throughout the district, while in some parts of the district, parents themselves provided private tutoring. The focus was on success in learning.

Finally, Loveless (2008) urged school districts to collect data and conduct their own research, so that they can examine the results of what they are doing. This step is critically important, so that light can be shed on latent structures that are feeding inequity. This district worked with an outside evaluator in the early stages of the reform. Not satisfied that the composition of their eighth-grade algebra classes was skewed in composition, they asked the evaluator to examine their existing mathematics enrollment data. As that evaluator, I found that it was not likely that Black students were in eighth-grade algebra, and that students' seventh-grade math class was a strong predictor of placement in eighth-grade algebra. However, selection for seventh-grade algebra was based on teacher nomination, tracking back as far as third grade. Table 5.2 contains the results of that analysis.

Moreover, the students' scores on the Stanford Achievment Test (SAT 9) were also indicated as a weak predictor of selection for eighth-grade algebra. Further investigation revealed that the SAT 9 scores were not a reliable predictor because students with identical SAT 9 scores were not guaranteed placement in eighth-grade algebra classes. What made the difference as to whether a student was in eighth-grade algebra? Teacher nomination sometimes removed students with acceptable scores. Readiness to learn, often re-

TABLE 5.2 Composition of Grade 8 Algebra by Prior Performance, Parent Education, Ethnicity: Predicting Performance by Algebra Group

Algebra: Grade 8	Odds Ratio	Std. Err.	*z*	*P* > [*z*]	95% Conf. Interval
Grade 7 Local Math grade	3.653107	.3125856	15.19	0.000	3.094994
Stanford Math Grade 8 Pretest	1.064388	.0030433	21.82	0.000	1.058439
Father's schooling	1.03174	.0132724	2.43	0.015	1.006052
Mother's schooling	.9932909	.0132272	–0.51	0.613	0.9677015
Identified Gifted	5.733845	2.192636	4.57	0.000	2.709845
Black/not Hispanic	.875644	.3507834	–0.33	0.740	.399335
Hispanic	1.181893	.1937996	1.02	0.308	.8570444
Caucasian	1.462724	.7242696	0.77	0.442	.5542283

Note: Number of cases = 2,634; $R^2 = 0.5362$; $p \leq .05$

lated to self-discipline and study habits, were cited as reasons for this removal. On the other hand, faced with parental pressure, school administrators sometimes inserted students into eighth-grade algebra who did not have acceptable SAT 9 scores. The result was that these students, with appropriate support in class and after school, succeeded in eighth-grade algebra!

Armed with this information and their own conviction that leveling the playing field would change the complexion of their eighth-grade algebra classes, the school leaders moved forward with their plan to implement algebra for (almost) all students in eighth grade. This was clearly a situation wherein policymakers "got it right" about implementing research-based reform. Every step of the way, the school leaders asked good questions and took their findings into consideration in their planning over the 5 years that it took to implement the reform. As a result of these efforts, this district turned around mathematics achievement for its students. Whereas only 30.7% of the district's eighth-grade students were enrolled in algebra in 2004–2005, effectively 90% of the eighth graders were studying algebra in the 2008–2009 school year.

Moreover, in this increased pool of eighth-grade algebra course-takers, 92% passed the state's Algebra 1 examination, scoring a bit lower than the state average (94%), which includes students studying algebra in ninth grade. Even more powerful are the results for various subgroups within the eighth-grade population that was studying Algebra 1. In every group, students scored comparably with the other test-takers in the state, even though the larger sample included students who were studying algebra in the ninth grade. The extra year of mathematics preparation did not seem to make an appreciable difference in the test scores.

Admittedly, the high passing scores on the state algebra test may indicate a larger problem in the rigor of the test. Nevertheless, the fact remains that on the state measure, the eighth-grade students in this district scored well when compared to others in the larger population across the state. As long as state tests are a universally accepted measure of student success, then these results are worth consideration. Moreover, district administrators are not content with simple passing scores. They have been following the "advanced pass rate" among their students taking the state Algebra 1 test, and that rate has also increased steadily since they implemented the eighth-grade algebra policy.

LESSONS LEARNED: RECOMMENDATIONS FOR FUTURE PRACTICE

The experiences of this large, diverse school district can serve as a template for action for other districts striving to address the complex issues surrounding detracking the delivery of the mathematics curriculum and increasing mathematics literacy across all populations. The paradigm that has been established leads to three clear recommendations.

Recommendation #1: Leaders must look at all available data and reexamine existing policies for inherently inequitable assumptions about the capacity of all students to study algebra in the eighth grade. Entrance qualifications and experiences leading to eighth-grade algebra deserve scrutiny, especially if they result in clear separation into tracks by race or ethnicity. Openness to change must come from leaders at all levels in the school administration, from the chief school officer through middle managers to curriculum coordinators and department chairs.

Recommendation #2: Provide support for both teachers and students in a *gradual* transition, starting with grade 5 at the latest, with attention to the teachers in grades 3 and 4 as well. The spiral curriculum in mathematics is incremental and recursive. Students' experiences in the elementary grades are critically important in providing an infrastructure of understanding for higher order tasks starting in grade 5. Following this district's model of success will require professional development for teachers at each grade level. Specific training sessions for teachers every 9 weeks in anticipation of the work of the next quarter will accomplish two goals. First, they will provide the teachers with concrete strategies and greater understanding of the mathematical concepts. Second, they will create the much desired "professional learning community" environment that has been proven to increase teacher effectiveness and efficacy.

Student support comes in two forms. First, schedule more time for mathematics in grades 6, 7, and 8. Time on task has long been regarded as a powerful tool for student comprehension. Second, provide additional tutoring and support sessions for students who struggle with the math concepts leading to and in the eighth-grade algebra course. The core belief is that every student can accomplish this goal. Some may need more time or support than others. At the same time, acknowledge that some students may well be ready for algebra before eighth grade. Once regarded as the elite prize for gifted students, the study of algebra should come even earlier for those who are mathematically talented (Assouline & Lupkowski-Shoplik, 2011).

Recommendation #3: Move forward deliberately and slowly. Achieving the goal of algebra for all students in eighth grade takes time. There are no easy fixes! The *Algebra for All* initiatives in Chicago and California failed because they were implemented too quickly, without laying a foundation of curriculum support in the grades leading to eighth grade and attaining teacher buy-in. The district in this study took 8 years to achieve the result that all students study algebra at least by eighth grade, regardless of who they are, where they live, and what temporal resources they might or might not have. They have achieved equity.

ACCESSIBILITY IS THE KEY TO EQUITY

Much has been said about achievement gaps in the United States, particularly in mathematics. On international test reports like the PISA, students in the United States score dismally in mathematics in comparison with their peers across the globe. In response to these reports, pundits frequently cite the problems inherent in teaching all students in a democracy and the unavoidable socioeconomic differences in resources among student populations, especially among those from lower socioeconomic environments. However, little has been said about the opportunity gaps that exist in some districts, particularly those with limited financial resources and low-income populations.

Raudenbush, Fotiu, and Cheoung (1998) long ago targeted the issue of access of minority students to the study of algebra and found that more affluent districts were more likely to provide algebra instruction to students in eighth grade than districts with fewer financial resources. Their results indicate "substantial evidence of inequality of access to these resources as a function of social background and ethnicity" (p. 265). Smith (1996) also found that "early algebra takers were less likely to be from minority groups and came from families of higher socioeconomic status" (p. 145).

Admittedly, students with fewer personal resources may have difficulty attaining success in school. Nevertheless, it remains the task of the school to address the needs of its student population in ways that are inclusive and foster success. Punitive, exclusive programming that makes it difficult, if not impossible, for students of diverse backgrounds to progress along a continuum of growth and success are simply inequitable. Tracking policies that sort students early in their academic careers are impediments to accessibility. Such policies do not allow for individual development and growth. Moreover, the success of the school reform in the district in this case study clearly shows that educators can provide smooth transitions in mathematics instruction for all students, as long as success is regarded as a series of incremental experiences rather than a set of hurdles or seemingly insurmountable steps.

Access to eighth-grade algebra can best be understood in the spirit of how architects now provide access to buildings in the United States. Whereas once all buildings were entered by means of stairs—steps, if you will—now there are ramps that lead to entrances and provide open access for all to approach the building under their own steam, with help if necessary. Architectural ramps now provide gradual access for all, from those who proceed quickly and race to the top to those who move more slowly, with accommodations like wheelchairs and walkers. No barriers, either physical or psychological, impede the individual's progress up the ramp.

This metaphor defines the success of this school district, in which access to eighth-grade algebra is no longer reserved for a select few. It is not reserved for those who can "make it to the top" of a series of steep steps. Instead, all students progress steadily up the ramp to algebra at least by eighth grade. Some get there sooner than others, but the vast majority get there! These school leaders accomplished substantial change in the school culture in their district. Educators at all levels collaborated to bring about this policy reform. Their efforts can serve as a template for other school leaders in other districts who strive to bring about realistic change in this tense era of accountability. The teachers are on board with the new plan and take pride in their success rates.

As a result of the policy reform in this district, the eighth-grade students in this district are succeeding on a state test that is usually taken by ninth-grade students in other parts of the state. While critics of state testing might maintain that state tests have been watered down to accommodate a larger testing pool, the fact remains that in this district, students are achieving this benchmark a full year earlier than their peers. Moreover, all students have the algebra credential in place, and many more are taking additional advanced math courses in high school than under the former policy. The students report greater college attendance. In the face of rhetoric about achievement gaps and "racing to the top," school leaders and the teachers in this district have closed the opportunity gap for students, at least going into high school, and continue to provide a strong foundation for further growth. That is no small accomplishment.

REFERENCES

American Educational Research Association (AERA). (2008, December). Special issue on foundations for success: The final report of the National Mathematics Advisory Panel. *Educational Researcher.* Thousand Oaks, CA: Sage.

Assouline, S., & Lupkowski-Shoplik, A. (2011). *Developing math talent.* Waco, TX: Prufrock.

Bass, H. (2006). *Presentation to the National Mathematics Panel,* Chapel Hill, NC. Retrieved from www.ed.gov/about/bdscomm/list/mathpanel/2nd-meeting/presentations/bass-hyman.doc

Boaler, J. (2006). Opening our ideas: How a detracked mathematics approach promoted respect, responsibility, and high achievement. *Theory into Practice, 45*(1), 40–46.

Campbell, G. (1996, May). Bridging the ethnic and gender gaps in engineering. *National Action Council for Minorities in Engineering Research Letter,* 6(11). Washington, DC: NACME.

Cavanaugh, S. (2008, July). Experts question California's algebra edict. *Education Week, 44.*

Cremin, L. (1980). *American education: The national experience.* New York, NY: Harper & Row.

Fennell, F. (2008, January/February). *What algebra? When? NCTM News Bulletin.* Reston, VA: NCTM.

Loveless, T. (2008). *The misplaced math student: Lost in eighth-grade algebra.* Washington, DC: Brookings Institution.

Moses, B., & Cobb, C. (2000). *Radical equations: Math, literacy, and civil rights.* Boston, MA: Beacon.

National Council of Teachers of Mathematics (NCTM). (2000). *Principles and standards for school mathematics.* Reston, VA: Author.

National Mathematics Advisory Panel. (2008). *Foundations for success: The final report of the National Mathematics Advisory Panel.* Washington, DC: U.S. Department of Education.

Pelavin, S., & Kane, M. (1990). *Changing the odds: Factors increasing access to college.* New York, NY: College Board.

Raudenbush, S. W., Fotiu, R. P., & Cheong, Y. F. (1998). Inequality of access to educational resources: A national report card for eighth-grade math. *Educational Evaluation and Policy Analysis, 20*(4), 253–267.

Smith, J. (1996). Does an extra year make any difference? The impact of early access to algebra on long-term gains in mathematics achievement. *Educational Evaluation and Policy Analysis, 18*(2), 141–153.

Spielhagen, F. (2006) Closing the achievement gap in math: Policy implications of eighth grade Algebra for All students. *American Secondary Education, 34*(3), 29–42.

Spielhagen, F. (2007). Closing the achievement gap in math: The long-term effects of eighth-grade algebra. *Journal for Advanced Academics, 18*(1), 34–59.

Spielhagen, F. (2011). *The algebra solution to mathematics reform: Completing the equation.* New York, NY: Teachers College Press.

Taylor, R. (1990). Teacher expectations of students enrolled in algebra. In E. Edwards (Ed.), *Algebra for everyone* (pp. 45–52). Reston, VA: National Council of Teachers of Mathematics.

CHAPTER 6

ADMINISTRATIVE SUPPORT OF STEM CULTURE

Brian Boyd and Matthew Grushon

INTRODUCTION

New schools focused on science, technology, engineering, and mathematics (STEM) have become more prevalent over the last decade. As districts and other groups consider starting a new school, or consider converting existing schools to a STEM model, they have a mandate to rethink what "school" looks like. If schools (STEM focused or otherwise) are to serve a different purpose, what structures and practices need to exist in these schools? How do school leaders go about developing those new sets of structures and practices?

This chapter will explore the experiences of one such STEM school over its first 3 years of existence. We will discuss four different tenets of the school that have contributed to its early success: equity and access to rigorous curriculum; project- and inquiry-based learning; teacher autonomy; and designing experiences for students with both rigorous content and qualities that transcend any one discipline, such as communication, creativity, or persistence. Throughout each section, we offer recommendations to help school leaders promote these types of activities through formal structures

Improving Urban Schools, pages 85–97

of the school, as well as more informal actions that help develop a culture among the adults that carry out the important work in schools.

Equity and Access to Rigorous Curriculum

Traditionally, only select students who "do school well" get access to a rigorous curriculum during the middle school and high school years. At the school level, there are many selective schools—private schools and other STEM or magnet academies across the country—that have admissions requirements for students to demonstrate some success in school prior to being admitted. Within schools, research has shown that students in honors and advanced-placement classes tend to be underrepresented as compared to students who are economically disadvantaged, students of color, and students with disabilities, to name just a few categories (Oakes & Lipton, 1994). Students are given access to rigorous curriculum often due to their competence in "doing school well" instead of their ability to perform well in a particular content area. For example, mathematics students who turn in their homework regularly and know how to prepare for procedure- and skills-based tests are placed in the higher level math classes (Horn, 2006). Decisions about whether students will take these courses in mathematics typically begin in middle school, when students are tracked into advanced courses. Therefore, decisions about students' access to rigorous mathematics courses are made early in their academic careers.

The Dayton Regional STEM School (DRSS) attempts to redefine this paradigm. DRSS is one of 10 public STEM-focused schools across Ohio offering students a relevant, real-world educational experience that will prepare them for college and opportunities in the work world. The school opened in August 2009 with its first class of ninth-grade students. Following two annual expansions, the school currently serves 320 students in grades 7–11 coming from 28 school districts across six counties.

Admissions acceptance at DRSS is not based on prior academic achievement. As a school of choice, students do have to submit a basic application to be entered into the admissions lottery, which is not impacted by prior test scores or grades. There are critics who believe that acceptance at a school such as ours must be merit based. We would argue that it is merit based: *all* students deserve access to a rigorous curriculum, especially those who may have been disenfranchised by a typical school experience in their past or may have circumstances that have prevented them from playing the "game of school" well prior to applying at DRSS.

DRSS was designed to reflect regional demographic indicators (poverty level, racial identification, special education) in order to prove that *all* students can learn and achieve when given an opportunity. In order to enroll a student body that is as diverse as the geographic region where the school is located, we had to recruit students and families in a way that would get their attention, interest them enough to apply, and hopefully spread the word to others. The recruitment effort was multipronged in its approach: radio spots, newspaper ads, special events (science fairs, festivals, among others), as well as informational sessions at numerous public libraries across the counties in the school's targeted region. We also reached out to school counselors who we felt would be able (and willing) to spread the word of the school to students they encountered, who, for one reason or another, were not finding success in their current situation.

In addition to accepting students from various academic backgrounds, at the classroom level, students are not tracked in any subject through the tenth grade. Students in grades 7–10 take a common set of courses. This means that regardless of what classes students take prior to attending DRSS, they have the same set of courses as every other student at the same grade level. This may seem counterintuitive to meeting the needs of individual students—placing them in the same set of classes regardless of their prior experiences. However, this allows for *all* students to have access to similar rigorous experiences. This also allows teachers to share a common set of students. In so doing, teachers are better able to collaborate and plan experiences that cut across disciplines, as well as beyond the walls of the school.

The curriculum at DRSS is set up so that teachers across content areas plan projects and units in which there is overlap between two content areas, at a minimum. During the first 2 years at the school, entire grade levels engaged in a at least two units that integrated *all* of the students' classes, including Mandarin Chinese, health and wellness, and fine arts. Teachers also collaborate on "smaller" units or projects that they determine as appropriate throughout the year.

While we have stated that all students take a common set of classes, it is important to note that this includes students with disabilities. Approximately 10% of DRSS students have an identified disability and an accompanying Individualized Education Plan (IEP). DRSS is a full-inclusion school when it comes to providing special-education services. There are no resource-room classes for individual subjects, and the intervention specialists employed by the school are tasked with keeping up with their students in each classroom as well as providing the appropriate (required) accommodations for their students. This adds a level of complexity for general-education teachers and intervention specialists alike, reflecting the commitment to meeting students at their level and giving them access to the same rigor as all other students.

This approach to a rigorous curriculum for *all* students can be difficult when marketing and promoting such a school to prospective students and parents. While we want to represent a picture of the work being challenging and rigorous, we do not want to alienate students who think that the school would be too difficult. It is important that students and their families understand that if they are willing to commit and work hard at their education, then the school will meet them at their level and work collaboratively toward their success. Our initial observations are that this approach has its challenges in the classroom as well. While philosophically the teachers are in agreement with our approach, the academically diverse set of students provides challenges that teachers must be willing to meet. In addition, the administration must be willing to provide structures and supports to enable them to do this. The commitment to including students from all backgrounds is incredibly important and impacts decisions that are made in other areas of the school administration.

Recommendations

Providing access to a rigorous curriculum in these ways can present a challenge for many stakeholders of a school. Typically, high school teachers are not trained to think this way about teaching their own subject, and they rarely have models that show how this could happen. We also rarely find high schools that provide a structure for teachers to make this possible. Having a common set of students, as well as a common planning time during the regular work day, is an example of a school structure that promotes this type of access to a rich and rigorous curriculum for all students. As a school leader, one must find the right balance of support and pressure for teachers to work in this capacity. Teachers will need to feel supported in taking risks to challenge students and in finding unique ways to support students in finding success within a rigorous class. As a leader, it is important to hold fast to this issue of access and to support teachers in making that access a reality.

Because project-based learning is an important part of this specific school, the issue of not tracking students has been very much a tenet we have maintained. One way to think about tracking is that you can choose to deal with issues that come with tracking or with the issues that come with not tracking (R. Riordan, personal communication, 2009). For reasons listed earlier in this section, we have chosen the latter approach. This will continue to be a topic that is debated and contested by students, parents, and teachers. These various stakeholders all enter the school with beliefs and previous experiences around how classrooms and schools should be structured to challenge students to best prepare them for their future. Trying to shift this paradigm for some teachers, students, and parents has not been simple or without debate and discussion. As with all issues, a school leader must listen and assess how well this is implemented in one's school. While there may be a time when the school may abandon this specific approach, the issue of access for *all* students is important to the schools' mission and cannot be abandoned without drastically altering the mission and vision of the school.

Project- and Inquiry-Based Learning

Another aspect that makes the curriculum of this individual school unique is its focus on making learning meaningful and relevant to students. Often the work students do in school seems disconnected from work in the real world (Horn, 2006). Teachers at DRSS often engage students in real problems and issues that exist and have connections to the specific content of their individual courses. For example, one project that tenth-grade students have done is around transportation and the proposed 3C+D high-speed rail system for Ohio, including a potential stop in Dayton. Students examined the proposed system and train stop from various perspectives that related to their specific classes. In life science class, students went to the proposed stop site and completed an Environmental Site Assessment (ESA). In social studies class, students used Global Information System software to look at population density for the Dayton area to inform whether Dayton could sustain having a rail stop. In math class, students developed surveys to gather information about train usage. Along the way, partners of the school provided expert advice to students. A local company that specializes in ESA work came to the school to provide examples and insight into how to complete an ESA and allowed students to ask questions prior to completing their own ESA of the proposed stop. A statistician from a local university worked with students to design a survey that would gather the data students needed to make an informed decision. Local government and Chamber of Commerce representatives spoke to the group about other impacts the rail system and stop could have on the region and state. The culminating event for the project was a presentation by groups of students. Using the information from their research and activities in their classes, students made a presentation to community members either advocating for or cautioning against the rail system and a Dayton stop. This is one example of a project that teachers designed that cut across disciplines as well as addressed an important issue for our partners and community members.

While this type of project-based learning is not a new concept, having this be a focus for an entire school is not overly common in U.S. schools. To engage students in this type of work, we decided to find opportunities in which students' work would be presented to, and assessed by, the public. This also ensures that the work serves a purpose beyond just the classroom or the purpose of satisfying a school requirement. It also raises the level of commitment of students. There is a real need for quality work they are proud of that will be viewed by others in addition to their teacher or classmates. If their work is going to be critiqued by experts in a certain field, students are more motivated to have the work be of high quality. This encourages the notion that our best work does not always happen the first time around, that it takes feedback and multiple drafts/iterations before it is our best work and something ready to be made public (Berger, 2003). It also fosters in our teachers a need for their classrooms to have structures for critiquing each other's work. This could be in the form of students giving each other feedback, teachers providing feedback, or experts in the field providing feedback. For example, when learning about various body systems in a biology class, students wrote part of a book chapter intended for middle school students on specific body systems. Students received feedback from second-year medical students as part of the critique process. For this and other types of feedback to be effective, teachers must have the right culture developed in their classrooms so that students do not perceive the feedback as punitive, but rather as a way to help make their work better.

One way that teachers develop these projects is to look across discipline lines to see what might be a natural connection. Teachers at DRSS are also encouraged to look for connections that are considered "outside" of the school walls. The 3C+D example involved using business and government partners to add relevance and expertise to the students' work. Today, with over 60 organizations and some 700 individuals who are actively involved in working with the school, these partners take a hands-on approach to make sure that DRSS students and educators are connected to the "real world." In addition to participating in specific projects, partners participate in curriculum development and review; job shadowing; "Power Lunch" presentations; career fairs; internship opportunities; student portfolio evaluation; debate judging; and various other in-kind supports. The importance of these partners helping to fulfill the school's mission of providing students with a rigorous education and grounded in the "real world" cannot be overstated.

Recommendations

While this work may sound exciting and sensible, it is not without its challenges. One common issue that a school leader must be prepared to consider is grading. There are often questions around what is graded as a part of a project and how teachers ensure that a grade reflects the content of a course and not just a student's ability to present a solution well to a group of guests to the school. While these are concerns that teachers must answer for their classrooms, it is also important to have some consistency or common thought about how these issues are handled. As parents and students experience this work, they sometimes have traditional expectations about grades and classroom structures.

Another issue to be aware of is the ambiguity that often exists around work on projects. Students, parents, and even teachers, are often accustomed to classrooms that are structured and in which solutions to classroom exercises are clear-cut with a straightforward path. True project-based learning engages students in various forms of inquiry, in which there usually is not an obvious path to a solution. There is typically a healthy amount of frustration among students and parents, as they have not been trained to think of school this way.

As a school leader, one must help teachers find the right balance between rigor in work that is truly like the real world and supporting students in ways so they do not become so frustrated that they shut down and become more turned off to school. The reality though is that for many students, project-based learning will be harder initially, because it is a different way of thinking about schoolwork. Students have been programmed to think about school and classes as a way to answer questions that have a predetermined right answer. Working on projects as described here involves a lot of ambiguity and difficult work in which there are multiple correct answers and solution paths. Some things that help students navigate this can be working collaboratively on large projects, as well as teachers providing checkpoints and targeted feedback about the type and level of work that is expected. In addition, the grading policy of the school also helps keep students persistent and engaged in the work. Certain projects and assignments can be given a score of "I," which means it is in progress and is not finished until it meets a higher quality. While the details of this can get messy, this communicates to students that they are capable of better work and that usually their best work does not happen the first time around (Berger, 2003).

Teacher Autonomy

Current policies and practices seem to take more decisions away from teachers. What should I teach next? How should I help students understand the structure and function of specific cells? The answers to these and many other similar questions are suggested for teachers, and in some districts, pacing guides dictate what is taught each day. While this is an attempt to control and ensure equity for students, it also tends to strip teachers of any autonomy or critical thinking about their work with students. In reality, regardless of the amount of content dictated to teachers, inside the classroom we cannot control most of the decisions teachers make—the questions they ask, how they respond to a student's question or an incorrect answer, how they motivate a student, or how they might inspire work that goes beyond the given state indicator. Teachers make hundreds of decisions during the course of a school day, regardless of how scripted some lessons have become: Should I answer that student's question with a more open-ended question? Which student's thinking should I highlight to help summarize today's lesson? Should I punish James for speaking out of turn even though he offered some of the best thinking on the topic today?

While the current push in education reform may be to deprofessionalize teaching, this is an area that is critical to any good teaching. Consider any award-winning teacher: Are they the ones who followed the pacing guide religiously? Are they the ones who avoided student questions or decontextualized the needs of their students to push ahead with the standards regardless of their students' needs, or are these winning teachers the ones who blaze their own path?

Do not be mistaken; we are not proposing that teachers swim against the current just for the sake of being autonomous. However, if as a school leader you want teachers to inspire students, to know students well, and find ways to challenge and support them, restricting teachers with rigid pacing guides and other authoritative structures will not lead to your intended outcomes. While teacher autonomy is needed, teachers still need to be responsive to their students' needs; the work they engage students in should still connect to work outside of their discipline and be seen as necessary outside the scope of school (as described in the previous section). But in order for teachers to accomplish this, they must be empowered and entrusted to make decisions about how that works. No pacing guide or administrative mandate will force that. Those decisions must come from the teachers who plan and implement that work with students.

At face value, this could seem as though we are promoting the typical high school teacher living in a bubble of his or her own discipline. Giving them autonomy does not mean you have to say "Do what you want." There are still constraints that we all must work within. At our school, those constraints include (a) planning what you think will be engaging and relevant to students, (b) finding as many ways over the year to connect students' work to another discipline and/or to the work of one of our partners, and (c) finding ways that students' work can be made public within some of the projects you plan.

The 3C+D project described earlier is one example of a project that would not have been implemented without teacher autonomy and the three guidelines above. Teachers had the freedom and support to investigate a topic that was both relevant and timely for our community and state. They worked tirelessly to plan and construct the project in a way that answered the guiding question of "What impact would the 3C + D Ohio Rail System have in the Dayton region?" and had students learning new mathematics, science, social studies and language arts content to help them answer this question. Teachers also provided a way for students to make their work public by having students present thoughtful, coherent information to community members about a pressing topic. No pacing guide or mandate from a curriculum office could have accomplished even a small fraction of what this project accomplished.

Recommendations

As the school leader who often feels a need to dictate such curriculum and instruction decisions, how does our role shift in a school in which teachers are encouraged to make these decisions for their classrooms? First, we still have to engage teachers in discussions about those decisions; they often need support and advice about those decisions. They also need to know that they can take a risk, that it's okay to do something and have it fail. Never is the entire project a failure, but there are always parts that can be improved. If teachers wait to have it all figured out, they will never implement something that pushes students, or themselves, out of their comfort zone.

In addition to more informal discussions, it is important to create formal structures that promote the culture of teachers discussing this work among themselves. As school leaders, we can be limited in the help we can provide as just one person (Glickman, 2002). Teachers need time to share their ideas about projects and get feedback from one another about the quality of the work. One structure to promote this is common planning time for a set of teachers who share a common set of students. Planning a project with multiple teachers takes an enormous amount of time and coordination. To do it well, we must provide teachers this time during their workday on a regular basis.

Another structure that can promote this type of collaborative work is having a group of teachers formally present and share a project that they have developed. This can be intimidating to teachers, especially ones who are used to being isolated in their approach to teaching their individual content. Having a basic protocol to follow for sharing these projects has proved helpful (Glickman, 2002). There are examples from Critical Friends and other networks that can be helpful for this type of work. Some basic categories of these protocols include a brief overview and goals of the project, some time for the audience to ask clarifying and probing questions, time for warm and cold feedback, and time for the presenting teachers to react to the questions and feedback. In addition, if student work is available, it is provided to consider as evidence of whether the project accomplished its goals. These sessions have provided at least a small venue for teachers to make their own work public, just as we expect our students to do. These sessions have also served as great tools for teachers to refine their own projects, and they have led to teachers seeking out each other's advice informally. Providing a time and structure for these types of discussions can help to promote the type of collaborative culture we want in the work of teachers and emphasize that teacher autonomy does not mean existing on an island.

Designing Experiences With Rigorous Content While Developing Qualities That Transcend School and Academic Disciplines

We are bombarded regularly with data that tell us our students are not doing well in mathematics and science. Historical and current NAEP results, as well as international comparisons from TIMSS and PISA scores, often tell us that U.S. schools are failing to produce the type of young adults we see as necessary for our country's future. While we could debate the validity of such comparisons, it is clear that our students need something different in terms of the STEM disciplines.

What is needed is not just *more* science and mathematics, but an approach to science and mathematics that is different. Doing science and mathematics well includes problem solving, critical thinking, reasoning, and communicating about significant mathematics and science concepts. Too often classrooms in mathematics and science work only to increase students' skills on isolated procedures—balancing a chemical equation, solving a quadratic equation, multiplying fractions—without students understanding the underlying concepts or how such skills might be relevant to the world outside of the school walls.

Key stakeholders to consider in this work are partners—business, governmental, or other community partners. These stakeholders have typically been considered exterior to the school, as educators have been reluctant to involve this group with their areas of expertise. At DRSS, we have found that engaging these stakeholders is extremely beneficial to our teachers and students. Businesses have opened their doors to our teachers and provide expertise and relevant contexts for students to engage. These partners have also been willing to serve in various capacities at the school: consulting with groups of students as they progress through a project, presenting to a group of students on an area of expertise that relates to specific content, providing feedback to students on their digital portfolios or other public work, and sharing information about their chosen profession at annual career fairs held by the school. These partners have also provided sites for job shadowing and internship opportunities for students. The partners of the school have truly enhanced the education provided by the school.

As we planned for the opening of our new school, we spent time asking partners what was important for them when they look to hire a new employee. What we heard was not just that they needed more employees with science, mathematics, or engineering backgrounds, but that their employees needed other skills as well: qualities such as communication skills, critical thinking skills, and the ability to work together with colleagues. As we visited other schools across the country, we also paid attention to how these types of qualities were identified and addressed in each school. All of these discussions informed our answers to the question: What do we want to be said about graduates of our school. As we discussed, debated, identified, and defined our focus, the qualities of persistence, inquiry, communication, creativity, and collaboration surfaced as important to us at DRSS. Regardless of what classes students are taking or what profession they anticipate entering, we believe that these qualities are important. These types of qualities were also consistent with what we saw at other STEM-focused schools and with what our business partners shared as important to them as employers. We have already mentioned some of their importance for doing science and mathematics well, but these types of qualities transcend just these disciplines. For example, being able to understand history well should also lead you to question (inquiry) whether events of the past might be playing out similarly today.

So while we want to increase the skills of students within individual disciplines that are being reemphasized in the educational landscape, there is equally a push to increase students' capacity for what some might call "soft skills" or "21st-century skills." Although we would argue that these are neither soft nor specific to the 21st century, we are glad they are a part of the focus. Working toward qualities such as persistence, inquiry, communication, creativity, and collaboration is not always easy or clear. We took time to discuss, debate, and finally identify, define, and describe these qualities in some detail. They are not perfect, they are just five "things." The next steps are equally, if not more, important: finding ways to make these explicit for students and deciding how we assess and provide feedback to students about their progress toward developing these qualities. How does a school place value on such things that are much harder to measure than a student's ability to solve a linear equation? How does one get teachers and students to value these types of qualities and then actually teach and incorporate them into the cultures of classrooms and an entire school? In order to really highlight these qualities and truly make them part of the culture of the school, we felt it necessary to assess these qualities explicitly. Otherwise, there was a risk that even though these qualities would be discussed, students and teachers would never focus their work directly on these qualities.

One way that we chose to address this at DRSS was through digital portfolios. Each year, students select work that they have completed through each of their courses that they are proud of and identify which quality was developed through that experience. Students complete a reflection about each experience that explains how the quality was demonstrated through that experience. In addition to an advisor evaluating the portfolio, business and community partners also assess how well students have progressed toward these qualities.

While these qualities transcend school and typical academic disciplines, students need to have a solid foundation within these typical school disciplines. Teachers must identify what is rigorous content within their set of courses. As we progressed through our first 3 years, we asked teachers to identify learning targets for each of their courses: What is important that students should be proficient in by the time they leave your classroom? At DRSS, we believe that these learning targets should be informed by state and national standards, but should also reflect the teachers' passion, work that is relevant and meaningful to kids and goes beyond any narrow state standard. Learning targets at one grade level should also be informed by what that student can expect to encounter in subsequent years. It is important for staff to share what they believe is important with one another and to help inform one another about the rigor they are trying to create in their own classrooms. These learning targets, however, should be content specific and not general. So while communication may be embedded, for example, in the eighth-grade mathematics teacher's learning targets, those targets should clearly identify mathematics content that is rigorous and leads to deep understanding.

Recommendations

As a school leader, promoting this type of work can seem to go against the grain of what is promoted in the current standards and accountability discussions. While school leaders and teachers have to embrace accountability, the type of accountability usually discussed boils down to students selecting the right multiple-choice answer. The accountability that we are talking about is a different sort of rigor. There may be overlaps with creating good test takers; but focusing only on test taking will not get you to rigorous content or the other qualities you want from your graduates (or teachers). As a school leader, deciding on what type of qualities your school will focus on is important. A school leader will need to lead and listen to discussions with key stakeholders to identify what skills make people successful in life. A school leader must be able to foster these discussions and reach collective decisions around the focus of a school so that teachers and support staff can appropriately work toward this focus.

Other important work in this area is the development and articulation of learning targets within a course. Helping teachers to develop these learning targets and providing ways for teachers to share and inform one another's targets is very important. While state standards might be dictated to us, teachers' decisions in the classroom have a major impact on which standards and targets are more thoroughly developed and which are just checked off a list. Developing clear learning targets and having colleagues to help those evolve is one way to help teachers explicitly communicate the decisions that they make within their classrooms. Promoting this type of transparency around one another's expectations is important in promoting the type of rigor we want in our schools.

CONCLUSIONS

The task of creating a school, whether from scratch or a conversion, should be one of the most exciting opportunities that a school leader is given. But, as you can see from the numerous questions posed, it is one that comes with as many challenges as opportunities. We frequently celebrated (and cursed) what we considered a once-in-a-lifetime opportunity.

If we were somehow granted a second chance to start a school, it would look a lot like what DRSS looked like in its first 3 years. We would also continue to work through the four tenets we highlight in this chapter. We have discussed the different tenets of the school that we feel are important and have contributed to our school's early success: equity and access to rigorous curriculum, project- and inquiry-based learning, teacher autonomy, and designing experiences for students with both rigorous content and transcendent qualities. While we have attempted to describe these in detail individually, it is clear that these areas are strongly interconnected.

These may not be the tenets other schools choose to focus on, but we believe that other school leaders will encounter many of the same issues as administrative practices are developed or modified in other school settings. Likewise, the circumstances that exist in each school are unique and therefore require an adaptation of what worked for us in the first 3 years at DRSS.

However difficult it may be to begin this work, we are confident that the payoff is worth it when considering the impact that a school will have on individual students, families, teachers, and communities at-large. Our focus is not on playing the "game" of school better than anyone else, but on putting in place the types of policies, practices, and supports that allow the "game" to be stripped away and the focus to be on educating students in a manner that fosters their development into critical thinkers, active and engaged citizens, and professionals of the future.

REFERENCES

Berger, R. (2003). *An ethic of excellence: Building a culture of craftsmanship with students.* Portsmouth, NH: Heinemann.

Glickman, C. (2002). *Leadership for learning.* Alexandria, VA: Association for Supervision and Curriculum Development.

Horn, I. S. (2006). Lessons learned from detracked mathematics departments. *Theory into Practice, 45*(1), 72–81.

Oakes, J., & Lipton, M. (1994). Tracking and ability grouping. In J. Goodlad & G. Keating (Eds.), *Access to knowledge* (pp. 187–204) New York, NY: College Entrance Examination Board.

CHAPTER 7

RACIAL AND LINGUISTIC ACHIEVEMENT DISPARITIES

Mathematics Course-Taking in Urban School Contexts

Eduardo Mosqueda
Saúl I. Maldonado
University of California, Santa Cruz

ABSTRACT

This chapter analyzes nationally representative quantitative data from the second (2004) wave of the Education Longitudinal Study (ELS: 2002–04) to examine the relationship between, schools' urbanicity, teachers' preparation, students' racial-ethnic identification, degrees of English-language proficiency (ELP), mathematics course-taking measures and twelfth-grade mathematics achievement. Highlighting relevant literature on urban education achievement disparities and employing a hierarchical linear model (HLM) method to account for students' clustering in specific schools, this chapter advances the argument that the effects of urban school contexts and students' characteristics (primarily race-ethnicity), ELP, and course-taking opportunities are complex, systematic, and significant. Our findings

Improving Urban Schools, pages 99–124

revealed that while the concentration of low-income students within schools has a negative effect on achievement, higher proportions of students enrolled in college-preparatory courses attenuates that effect. Additionally, course-taking resulted as the most influential factor on achievement. Further, while ELP had a negative effect on the achievement of Latino and Asian linguistic minority students (LMs), relative to non-LMs, the effect of ELP on achievement was more pronounced for Latinos. Beyond increasing opportunities to learn, such outcomes suggest a reconsideration of urbanicity not only as an indicator of achievement disparities but also as a marker of specific pedagogical opportunities that have the potential to redress inequitable outcomes associated with urban schools.

INTRODUCTION

Relationships among students' SES, race-ethnicity, English-language proficiency, and mathematics achievement have been persistent features of educational research in the United States (Gándara & Contreras, 2009; Secada, 1992, 1996; Smith, 1995; Tate, 1997). Consistently, mathematics achievement disparities are representative of broader patterns of social inequality—White, Asian, affluent, and English-proficient students outperform Latinos, Blacks, poor, and LMs. In urban school contexts, such academic achievement disparities are often unquestioned. In order to develop viable solutions to equalize the STEM attainment of urban high school students, a critical examination of the intersection of equitable access to advanced mathematics coursework and mathematics achievement is required.

In this chapter, we argue that the mathematics achievement of urban secondary school students is dependent upon several interrelated features. Our definition of urban school contexts considers the interplay of structural factors (e.g., race-ethnicity, poverty) and institutional mechanisms (e.g., access to both rigorous content and adequately prepared teachers). Thus, our examination of secondary schools explicitly investigates the urban and instructional contexts that differentially influence students' access to advanced mathematics education. By urban school contexts, we primarily refer to schools' socioeconomic status (SES) and racial/ethnic-linguistic composition and concentration. By instructional contexts, we principally refer to student's access to two specific resources: rigorous content and prepared teachers. We hypothesize that a robust understanding of the interrelatedness between instructional and urban contexts in secondary schools has the potential to redress mathematics achievement disparities.

BACKGROUND

The mathematics underachievement of urban school students is a cause for national concern. On average, urban school students underperform relative to suburban and rural students (Tate, 1997). Noguera (2003) has argued such underperformance is explained by the relationship among "compositions of neighborhoods and schools, school location and the test scores of children" (p. 43). Further, recent patterns of immigration have resulted in shifting neighborhood composition and school concentration of LMs who speak a language other than English at home. Nearly 11.2 million 5- to 17-year-olds (21% of the national total) were LMs (NCES, 2011) in 2009. Often unfamiliar with schools and school systems in the United States, LMs are more likely to live in poverty and matriculate in urban schools with content knowledge mastery below their language majority counterparts (Abedi & Gándara, 2006; Gándara & Contreras, 2009; Ruiz-de-Velasco & Fix, 2000). Considering the concentration of LMs in urban schools and their identification as the most underperforming subpopulation on standardized exams (Abedi, 2004; Gándara & Contreras, 2009; Mosqueda, 2010; Secada, 1992), current research must analyze achievement as influenced by urban contexts.

That stated, previous explanations of urban students' mathematics underachievement have overemphasized student demographics (e.g., SES, racial-ethnic segregation, degree of English-language proficiency) as central factors inhibiting academic success. For instance, it has been argued that SES and racial-ethnic segregation greatly impact student achievement (Orfield & Lee, 2005; Rumberger & Palardy, 2005). We believe that accounting for schools' demographic contexts is important but insufficient as explanatory indicators for mathematics achievement and postsecondary STEM opportunities.

We contend that an overemphasis on demographic descriptors distorts how we understand urban students' mathematics underachievement because it turns our attention from the more important pedagogical and institutional sources of inequity. Therefore, redressing mathematics disparities in urban educational contexts must consider the nuanced complexities of aggregate achievement patterns, variability in students' English-language proficiency, and institutional mechanisms such as equitable access to rigorous college preparatory instruction.

Bearing in mind the importance of access to rigorous mathematics content in structuring academic achievement opportunities (Adelman, 2006), course-taking trends function as measurements of mathematical rigor. Further, course-taking indicators must also account for the inextricable association with teachers' preparation. Considering that previous research has advanced teacher preparation as an essential facilitator of educational quality and equity (Darling-Hammond, Berry, & Thoreson, 2001; Darling-Hammond & Youngs, 2002; Gándara & Maxwell-Jolly, 2006), urban schools' research must examine academic achievement trends in relation to students' access to rigorous content and adequately prepared teachers (Capraro, Capraro, Yetkiner, Rangel-Chavez, & Lewis, 2010; Capraro, Young, Lewis, Yetkiner, & Woods, 2009).

The purpose of this study is to explore the synergistic relationship between demographic composition, instructional contexts, and mathematics achievement in urban schools. This chapter utilizes two-level hierarchical linear models (HLM) to investigate the effect of individual-level characteristics of students (Level 1) and school-level characteristics (Level 2) on twelfth-grade mathematics achievement in the Education Longitudinal Study of 2002–04. The sample analyzed contains 12,738 students from 748 schools, a nationally representative sample of Asian, Black, Latino, and White high school seniors in U.S. schools in 2004. Our inquiry is guided by the following research questions:

Does access to advanced courses influence the mathematics performance of Asian, Black, Latino, and White secondary school students in urban school contexts at the end of high school?

Does access to advanced courses influence the mathematics performance of linguistic minority Asian and Latino students relative to Asian, Black, Latino, and White non-LMs in urban schools at the end of high school?

Urban School Contexts and Achievement Disparities

Urban schools educate the most vulnerable and underserved students. Often located in communities that are socially and economically depressed and isolated, such schools mirror the realities of urban communities, where needs greatly surpass the available resources (Kozol, 1991; Maerrof, 1988; Noguera, 1996; Wilson, 1987). Noguera (1996) has argued that in such marginalized communities, "urban public schools [are] one of the few social institutions that remain more-or-less intact and that continue to provide a degree of stability" (p. 4). As such, urban schools are a potential source of hope for expanding postsecondary school opportunities and job prospects that, in turn, may help revitalize urban areas. Given the inextricable link between urban schools and racial-ethnic and income segregation in U.S. communities, it is within these vulnerable and inequitable settings wherein a majority of Latino and Black students receive their educational preparation (Gándara & Orfield, 2010; Noguera, 2003; Oakes, 1985; Valencia, 2002). Thus, efforts to increase the number of students attending urban schools in STEM-related fields must first look beyond the "milieu of hopelessness and despair" (Kretovics & Nussel, 1994, p. ix) that reflects economic inequality in urban areas and delve into the instructional possibilities that can maximize academic potential. Garnering additional resources to confront inequalities for urban schools is no simple task, considering Noguera's argument that underachievement in such schools is often socially positioned as the exclusive responsibility of urban communities, parents, and students.

Previous studies examining the link between urban school contexts and academic achievement have focused on the within-school segregation of students by race-ethnicity and/or income (Orfield & Lee, 2005). A recent study of the effects of school segregation, a central indicator of a school's urbanicity, found that the greatest impact on academic achievement was school's socioeconomic composition and not racial-ethnic composition (Rumberger & Palardy, 2005). Such findings suggest that students who attend schools where average students' SES is affluent (those with the highest socioeconomic composition) received the greatest academic benefits, while students attending schools where average students' SES is low showed depressed achievement outcomes.

While student-level and school-level SES are significant moderators of academic achievement, racial-ethnic and linguistic realities of students in urban school contexts cannot be ignored. For instance, the average mathematical achievement of twelfth-grade Black and Latino students is comparable to the average mathematical achievement of eighth-grade White students (Wilkins & Educational Trust, 2006). Further, achievement disparities in urban schools must also consider racial-ethnic and linguistic segregation. Research on school segregation has focused on the inequitable opportunities to learn[1] and unacceptably low outcomes in schools that primarily serve racial-ethnic minority students (Gándara & Orfield, 2010; Valencia, 2002). In addition, research has begun to examine the role of linguistic segregation. For instance, Gándara and Orfield (2010) found that the excessive linguistic segregation of Latino LMs had negative consequences on academic achievement as well as social and emotional development. Relatedly, Gifford and Valdés (2006) found that Latino LMs were concentrated in the lowest performing schools and that such linguistic segregation influenced both the academic achievement of LMs and opportunities to learn basic and academic English.

English Language Proficiency and Mathematical Language

Extant literature on demographic contexts of urban schools has underemphasized transformations in racial-ethnic student enrollment patterns. As of the 2005–2006 academic year, public schools' enrollment of Asian and Latinos had surpassed Blacks as most likely to attend urban schools in the largest cities (Frankenberg, 2009; Orfield & Lee, 2005). For Latino and Asian LMs, considerations of academic achievement must explicitly account for ELP.

Given the influential role of ELP on students' academic achievement, it is important to consider the variability in both English and native-language proficiency. However, the designation of LMs does not adequately capture students' extensive range of native language; some exemplify grade-level mastery in their native language while others may enter schools with underdeveloped language skills (Ruiz-de-Velasco & Fix, 2000). Such varying levels of Asian and Latino students' ELP and native language play an important role in learning mathematics, specifically because of the complexity of rigorous secondary school mathematics content (August & Hakuta, 1997; Khisty, 1995). As a complex register of words, expressions, and meanings (Cuevas, 1984; Secada, 1992), mathematics is itself a language that appropriates everyday English words and phrases (Garrison & Mora, 1999; Gutiérrez, 2002) and as such, can have important implications for assessment.

Given that local, state, and national standardized assessments in mathematics are exclusively administered in English (Dentler & Hafner, 1997; Gutiérrez, 2002, 2007; Lucas, Henze, & Donato, 1990), mathematical language intersects with ELP. Therefore, LMs are simultaneously evaluated on their comprehension of abstract and complex mathematical concepts *and* their ELP. This assessment approach raises uncertainties as to whether the assessment measurements are more indicative of mathematical content knowledge or English language and literacy conventions (American Educational Research Association, 1999; Boaler, 2008; Solano-Flores, 2011). For instance, in the following California High School Exit Exam (CAHSEE) statement, "In probability, an event is a particular happening that may or may not occur," the word *event* has a specialized meaning, the event is a conditional *particular happening*, and a relative clause modifies *happening* (Wong-Fillmore, 2007, p. 339). Only those test takers with high degrees of ELP are likely to not be confused by nonmathematical language and answer the item correctly. Therefore, research that examines the academic achievement (as measured by standardized tests) patterns of Asian and Latino students in urban school contexts must explicitly account for the influence of ELP.

Course-Taking Opportunities and Teacher Qualifications

In addition to demographic contexts of urban schools, mathematics underachievement must also consider pedagogical and institutional sources of inequity. Research has shown that attending urban schools is linked to limited access and opportunities to take rigorous content courses (Oakes,

1985; Oakes, Rogers, Silver, Horng, & Goode, 2004). Thus, an important influence in urban school achievement is the inequitable access to college preparatory instruction (Oakes, 1985). Curricular choices, particularly in mathematics and science, are grossly limited compared to schools' course offerings in nonurban contexts (Oakes et al., 2004). Further, urban school classrooms are often disproportionally likely to have inadequate instructional materials as well as an overrepresentation of teachers without the appropriate qualifications or credentials (Gándara & Contreras, 2009; Oakes et al., 2004; Olsen, 1995).

The relationship between mathematics course-taking patterns and higher standardized test scores (Callahan, 2005; Mosqueda, 2010; Wang & Goldschmidt, 1999) and future educational opportunities (Pelavin & Kane, 1990) is not exclusive to students in urban schools. Studies have shown that secondary school students who completed mathematics courses beyond Algebra 2 (e.g., Trigonometry, Pre-Calculus and Calculus) were twice as likely to earn a college degree compared to those who only took Algebra 2 and below (Adelman, 2006). Considering the importance of access to algebra and other gatekeeper curricula in structuring academic achievement opportunities (Secada, 1992, 1996; Smith, 1995; Tate, 1997; Wang & Goldschmidt, 1999), course-taking indicators must also consider the inextricable association with teachers' content-area professional qualifications.

Access to adequately prepared teachers is a critical issue for racial-ethnic and linguistic minority students since their coursework opportunities are disproportionally facilitated by teachers lacking appropriate content knowledge preparation (Flores, 2007; Maxwell-Jolly & Gándara, 2002). Not only are LMs linguistically isolated from non-LM peers and placed in courses taught by the least prepared teachers (Minicucci & Olsen, 1992; Valdés, 1998), their teachers' efficacy in both conveying content and providing linguistic supports is alarmingly low (August & Hakuta, 1997; Ruiz-de-Velazco & Fix, 2000; Téllez & Waxman, 2006). In addition, LMs are disproportionally underenrolled in college-preparatory coursework (Callahan, 2005; Olsen, 1995), specifically in mathematics (Mosqueda, 2010). Further, even when LMs are reclassified as English-language proficient, such classification does not guarantee students' access to rigorous courses (Olsen, 1997; Valenzuela, 1999) in mathematics (Wang & Goldschmidt, 1999).

Research has shown course-taking opportunities are significant indicators of academic achievement for students in general (Adelman, 1999, 2006; Lee & Bryk, 1988; Lee, Croninger, & Smith, 1997) and for LMs in particular (Callahan, Wilkinson, & Muller, 2010; Callahan, Wilkinson, Muller, & Frisco, 2009; Wang & Goldschmidt, 1999). In concert with the detrimental impact of underprepared teachers (Hanushek, 1992) and the academic achievement of urban school students, researchers must analyze *opportunities to learn*, which consider the interplay of access to both rigorous course-taking opportunities as well as to prepared teachers.

METHOD

The data for this study are drawn from the second wave of the Educational Longitudinal Study (ELS, 2004), a large nationally representative dataset provided by the National Center for Education Statistics (NCES). ELS provides policy-relevant trend data about critical transitions experienced by a national probability sample of students as they proceed through high school and into college or their careers (Ingels, Pratt, Rogers, Siegel, & Stutts, 2004). The ELS dataset is composed of students who were in the tenth grade in 2002 and in twelfth grade in 2004. The complete ELS sample includes 15,362 students from a random sample of 752 public, Catholic, and other private schools. The dataset contains assessments of students in reading and mathematics performance in addition to measures of important student, family, teacher, classroom, and school characteristics.

This analysis was based on the first follow-up (2004) subsample (weighted) of 12,738 students clustered in 748 schools. Statistical power analyses (Light, Singer, & Willett, 1990) suggest that this sample size provides power sufficient (.90) to detect small effects at the typical levels of statistical significance ($\alpha = 0.05$).

Variables in the Models

Outcome Variable

Twelfth-grade mathematics achievement ($F1MSTD_{ij}$) represented an IRT (Item Response Theory) scaled score (recoded from F1TXMSTD), which provides a norm-referenced measurement of achievement relative to the population of high school seniors in 2004 (Ingels et al., 2004) for each student *i* in school *j*. The ELS assessment itself contained items in arithmetic, algebra, geometry, data/probability, and advanced topics (Ingels et al., 2004). These scores were standardized to a mean of 50 and a standard deviation of 10 in the *full* ELS sample (Ingels et al., 2004). The mean test scores for the subsample of students in this study is 50.72, with a standard deviation of 10.12 points on the ELS assessment (ranging from 19.82 to 79.85). IRT scaled scores were used because they simplify the interpretation of predictors' impact on the outcome. A 1-point difference associated with the outcome variable equals one item correct on the ELS assessment. For a more detailed description of all of the variables included in this analysis, see Table 7.1.

TABLE 7.1 Descriptive Statistics for All Variables

Variable	Description	*n*	Mean	S.D.	Min.	Max.
Student Characteristics						
SES	ELS:2002 Standardized SES Composite	12,738	0.08	0.74	–2.12	1.87
FEMALE	Sex (0 = Male and 1 = female)	12,738	0.50	0.50	0	1
Race/Ethnicity		12,738				
ASIAN	Indicates whether the students self-identifies as Asian	12,738	0.11	0.31	0	1
BLACK	Indicates whether the students self-identifies as Black	12,738	0.13	0.34	0	1
LATINO	Indicates whether the students self-identifies as Latino	12,738	0.15	0.35	0	1
WHITE	Indicates whether the students self-identifies as White	12,738	0.61	0.49	0	1
English Proficiency		12,738				
LM	Nonnative English speaker (LM) (0 = Native English Speaker)	12,738	0.15	0.35	0	1
E_PROF_LM	Degree of English-Language Proficiency (ELP)	12,738	6.97	0.90	3	8
Course Placement		12,738				
HIGHMATH	Highest Mathematics	12,738	4.12	1.13	0	5
Teacher Preparation		12,738				
MTHMAJOR	Math teacher has degree in math	12,738	0.57	0.50	0	1
MTHCERT	Math teacher is certified/credentialed	12,738	0.77	0.42	0	1
School Context Measures						
PUBLIC	School Control (1= Public and 0 = Catholic or other private)	748	0.77	0.42	0	1
URBAN	Urbanicity (0 = Suburban or Rural and 1 = Urban)	748	0.33	0.47	0	1
SCHOOLSES (F_FLUNCH)	% of twelfth-grade students who qualify for free/reduced-price lunch	748	3.59	1.90	1	7
PCT_LEP (B_G10LEP)	% of twelfth graders who are LM students	748	3.17	1.87	0	4
PCT_COLTRK	% of twelfth graders enrolled in the college/academic track	748	6.30	2.04	1	10
PCT_CREDT	% of twelfth graders' teachers who are credentialed	748	9.25	1.85	1	10

Question Predictors

The first primary question predictor is $F1HIMTHR_{ij}$ and indicates the highest mathematics course taken by the end of each student's senior year of high school. Given that high school mathematics courses are typically taken in sequence, this variable is treated as an interval scale (0 = no math taken, 1 = Pre-Algebra or basic math, 2 = Algebra 1, 3 = Geometry, 4 = Algebra 2, and 5 = Trigonometry, Pre-Calculus, or Calculus).

The second primary question predictor is the dummy variable LM_{ij}, which is used to distinguish between LM and non-LM students (1 = LM, 0 = non-LM). About 15% of the students in this sample reported being LM. LM status was recoded from F1STLANG, a survey question that asked students, "Is English your native language (the first language you learned to speak when you were a child)" (Ingels et al., 2004). Additionally, in order to differentiate among the level of English proficiency of LM students, the cross-product $LM_{ij} \times E\_PRF\_LM_{ij}$ is used. $E\_PRF\_LM_{ij}$ is a composite that ranges from 3 to 8 (low to high), based on each student's self-reported level of English proficiency.[2] This weighted composite score is constituted from students' responses to four ordinal dimensions of self-reported English proficiency that include how well students "understand spoken English," "speak English," "read English," and "write English."[3] For each of these dimensions of English proficiency, students provided one of the following ordinal responses: "Very well," "Well," "Not well," or "Not at all." Descriptive statistics for the analytic sample are displayed in Table 7.2.

Control Predictors

The analysis includes a series of control predictors in order to account for individual background and school context variation that may influence the outcomes. These predictors are also useful to assess the potential impact of selectivity bias. These controls include student-level SES, gender, race-ethnicity, and degree of ELP, in addition to teacher-level preparation measures. At Level 2, controls are included for selected school contexts such as whether the school type is public or private, urban, and the following percentages: low-income students, LM students, students on the academic track, and credentialed teachers.

Sample Weights

The ELS student-level panel weights (F1QWT)[4] and school-level weights (BYSCHWT) were applied to the analysis according to the guidelines provided for the HLM software (Raudenbush & Bryk, 2002).

TABLE 7.2 Data and Coding for All Variables

Variable	Definition	Notes/Coding
Student Characteristics		
SES	ELS:2002 Standardized SES Composite	Ranges from –2.12 to 1.87
FEMALE	Students' Gender	1 = female, 0 = male
Race/Ethnicity		
ASIAN	Indicates whether the students self-identifies as Asian.	1 = yes, 0 = no
BLACK	Indicates whether the students self-identifies as Black.	1 = yes, 0 = no
LATINO	Indicates whether the students self-identifies as Latino.	1 = yes, 0 = no
WHITE	Indicates whether the students self-identifies as White.	1 = yes, 0 = no
English Proficiency		
LM	Nonnative English speaker (LM)	1 = nonnative, 0 = native speaker
E_PROF_LM	Degree of English-language proficiency (ELP)	Ranges from 3 to 8
Course Placement		
HIGHMATH	Indicates the most advanced mathematics course of each respondent.	0 = no math taken 1 = Pre-Algebra or basic math 2 = Algebra 1 3 = Geometry 4 = Algebra 2 5 = Trigonometry, Pre-Calculus, or Calculus
Teacher Preparation		
MTHMAJOR	Math teacher has degree in math	1 = yes, 0 = no
MTHCERT	Math teacher is certified/credentialed	1 = yes, 0 = no
School Context Measures		
PUBLIC	School control	1 = Public 0 = Catholic or other private
URBAN	Urbanicity	0 = Suburban or rural 1 = Urban
SCHOOLSES (F_FLUNCH)	% of twelfth-grade students who qualify for free/reduced-price lunch at the high school	Ranges from 1 to 10.
PCT_LEP	% of twelfth graders who are LM students in the high school	Ranges from 0 to 4.
PCT_COLTRK	% of twelfth graders who are enrolled in the college/academic track	Ranges from 1 to 10.
PCT_CREDT	% of twelfth graders' teachers who are credentialed	Ranges from 1 to 10.

ANALYTIC STRATEGY

To address the research questions, we computed three nested fitted models for twelfth-grade mathematics achievement outcomes. The first model in the analysis is the fully unconditional model (a random effects ANOVA model) that contains no independent (control) predictors and includes only the mathematics achievement outcome (F1MSTD_{ij}). In the unconditional model, the outcome variable is analyzed as a function of the intercept (β_0), which represents the institutional average of each student's twelfth-grade mathematics score, a random effect that is unique to each individual (r_{ij}) and a random effect that is unique to each school (u_j): $\text{F1MSTD} = \beta_0 + (r_{ij} + u_j)$.

The unconditional model also provides important information about the intraclass correlation (ICC) or the proportion of variance in each outcome that is attributable to the between-institution differences.[5] The estimated ICC in Model 1 shows that 20.4% of the variance in mathematics achievement outcomes is due to between-school differences, and the remaining 73.5% proportion of variance is due to within-school differences.[6] These between-institution differences in the proportion of variance in mathematics achievement scores are statistically significant and indicate the appropriateness for a multilevel modeling analytic strategy.

The second model in the analysis is our baseline model and contains the main effects of the student- and school-level control variables in the analysis and corresponds to the first research question. HLM was used to examine the main effect of how student- and school-level factors and course taking relate to the twelfth-grade mathematics achievement scores. The fitted multilevel regression models corresponding to the first research question were:

$$\begin{aligned}\text{F1MSTD}_{ij} = {} & \beta_0 + \beta_1\text{FEMALE}_{ij} + \beta_2\text{SESF1}_{ij} + \beta_3\text{MT\_CERTR}_{ij} + \\ & \beta_4\text{MT\_MJRIR}_{ij} + \beta_5\text{F1HIMTHR}_{ij} + \beta_6\text{LATINO}_{ij} + \beta_7\text{BLACK}_{ij} + \\ & \beta_8\text{ASIAN}_{ij} + \beta_9\text{PUBLIC}_j + \beta_{10}\text{URBAN}_j + \beta_{11}\text{SCHOOLSES}_j + \\ & \beta_{12}\text{PCT\_LEP}_j + \beta_{13}\text{PCT\_COLTRK}_j + \beta_{14}\text{PCT\_CREDT}_j + \\ & \gamma_1 Z_{ij}^{7} + \gamma_2 Z_j^{8} + (r_{ij} + u_j)\end{aligned}$$

The third model in the analysis addresses the second research question and tests for the differential impact of English proficiency and course-taking patterns on twelfth-grade achievement. The main effects between access to advanced mathematics courses and mathematics achievement were added to the multilevel model in the previous equation as follows:

$$\begin{aligned}\text{F1MSTD}_{ij} = {} & \beta_0 + \beta_1\text{FEMALE}_{ij} + \beta_2\text{SESF1}_{ij} + \beta_3\text{MT\_CERTR}_{ij} + \\ & \beta_4\text{MT\_MJRIR}_{ij} + \beta_5\text{F1HIMTHR}_{ij} + \beta_6\text{LATINO}_{ij} + \beta_7\text{BLACK}_{ij} + \\ & \beta_8\text{ASIAN}_{ij} + \beta_9\text{LM}_{ij} + \beta_{10}\text{E\_PRF\_LM}_{ij} + \beta_{11}\text{PUBLIC}_j + \\ & \beta_{12}\text{URBAN}_j + \beta_{13}\text{SCHOOLSES}_j + \beta_{14}\text{PCT\_LEP}_j + \beta_{15}\text{PCT\_} \\ & \text{COLTRK}_j + \beta_{16}\text{PCT\_CREDT}_j + \gamma_1 Z_{ij} + \gamma_2 Z_j + (r_{ij} + u_j)\end{aligned}$$

RESULTS

In our analysis, we examined the estimated effects of student characteristics, ELP, course taking, teacher preparation, and school context measures on the twelfth-grade mathematics achievement of Asian, Black, Latino, and White students. Model 1 in Table 7.3 (the unconditional model) shows that the intercept representing the average mathematics achievement in the population of all twelfth-grade students is 49.98 on the ELS: 2004 assessment.

In Model 2, we examined all of the aforementioned baseline control variables and found statistically significant relationships between twelfth-grade mathematics achievement, students' SES, gender and race-ethnicity, teacher preparation measures, and schools' percentages of SES and academic track students.

TABLE 7.3 HLM Fitted Models

	12th Grade Mathematics Achievement		
Fixed Effects	**Model 1 coef. (SE)**	**Model 2 coef. (SE)**	**Model 3 coef. (SE)**
Intercept	49.88***	50.35***	50.49***
	(0.29)	(0.48)	(0.48)
School Context Measures			
Public school		–0.87	–0.99
		(1.08)	(1.06)
Urban		0.76	0.85
		(0.60)	(0.60)
School SES (% of twelfth-grade free/reduced-price lunch)		–1.06***	–1.01***
		(0.15)	(0.14)
% of twelfth-grade LM		–0.19	0.44
		(0.32)	(0.30)
% of twelfth-grade college track		0.31***	0.30***
		(0.09)	(0.09)
% credentialed teachers		0.22	0.21
		(0.18)	(0.18)
Student Characteristics			
Socioeconomic status		2.34***	2.34***
		(0.16)	(0.16)
Female (male is omitted)		–1.70***	–1.70***
		(0.20)	(0.20)
Race-Ethnicity			
Asian		1.03~	1.12~
		(0.54)	(0.60)

(continued)

TABLE 7.3 HLM Fitted Models (continued)

	12th Grade Mathematics Achievement		
Fixed Effects	**Model 1 coef. (SE)**	**Model 2 coef. (SE)**	**Model 3 coef. (SE)**
Black		–4.60***	–4.62***
		(0.36)	(0.36)
Latino (White omitted)		–2.52***	–2.30***
		(0.40)	(0.44)
English Proficiency			
Linguistic minority (LM)		–2.13***	–1.67**
		(0.51)	(0.53)
Degree of English lang. prof. (ELP)		0.24**	0.15
		(0.09)	(0.11)
Course Placement			
Highest mathematics		4.01***	4.01***
		(0.16)	(0.16)
Teacher Preparation			
Math major		1.01***	1.01***
		(0.30)	(0.30)
Teacher certification/credential		1.17***	1.17***
		(0.35)	(0.36)
Interactions			
Linguistic minority (LM) × Latino			–4.17*
			(1.92)
English lang. prof. (ELP) × Latino			0.59*
			(0.28)
Random Effects			
Between schools ($\tau00$)	20.04	10.60	10.50
Within schools ($\sigma2$)	73.45	46.03	46.03
Chi-square	4413.87***	3737.43***	3702.13***

Key: ~ p < .10; * p < .05; ** p < .01; *** p < .001

The effects of race-ethnicity on mathematics achievement were similar and therefore not statistically significant between White students (the omitted category), who on average scored 50.35 on the ELS: 2004 assessment and the scores of Asian students ($\beta = 1.03$, $p > .05$). However, we did find statistically significant yet lower results between Whites and both Blacks ($\beta = -4.60$, $p < .001$) and Latinos ($\beta = -2.52$, $p < .001$). On average, Black students scored more than 4.5 points lower than White students, a difference of 45% of a standard deviation; Latino students scored 2.5 points lower than White students, a difference of 25% of a standard deviation; Asian students scored 1 point higher than White students, a difference of 10% of a standard deviation.

The parameter estimates also showed students' SES had a positive effect on mathematics outcomes. A one-unit of positive difference in SES was associated with over a 2-point higher difference on the ELS: 2004 assessment ($\beta = 2.34$, $p < .001$), a difference of over 20% of a standard deviation. The effect of gender, however, was negative. On average, the scores of females were near 1.5 points lower than the scores of males ($\beta = -1.70$, $p < .001$), a difference of 15% of a standard deviation.

In addition, Model 2 evaluates the effect of LM status, students' ELP (ENGPROF), along with the highest mathematics course taken at the end of high school on the ELS: 2004 mathematics assessment. As previously discussed, LM status and ELP were coded as cross-products ($LM_{ij} \times ENGPROF_{ij}$), therefore these effects cannot be interpreted independently. Our findings reveal that ELP is an important predictor of mathematics achievement. The results show that the aggregate test scores of LMs with low levels of ELP were below the aggregate test scores of their LM peers with higher levels of ELP. Thus, while on average, the scores of LMs were 2 points lower than non-LMs' achievement ($\beta = -2.13$, $p < .001$), every one-unit positive difference in ELP was also associated with nearly one quarter point positive difference in mathematics achievement on the ELS: 2004 standardized test ($\beta = 0.24$, $p < .05$), all other predictors being equal. This difference in performance is equal to one fortieth of a standard deviation *for every unit change in ELP*. For example, the combined effects of ELP and LM status show that LMs at the lowest level of English proficiency, on average, score more than one and one-half points (1.72) below the mean achievement of non-LMs, an important difference totaling more than 15% of a standard deviation, an effect that is reduced by one-quarter point for every unit change in ELP.

Notably, course taking resulted as the most important predictor of mathematics achievement. The results from Model 2 showed that taking higher-level mathematics courses by the end of high school had a positive and considerable effect on mathematics achievement. On average, *every* additional higher-level mathematics course in the high school sequence taken by individual students was associated with a 4-point positive difference in aggregate mathematics scores for all students ($\beta = 4.01$, $p < .001$), a relative change of 40% of a standard deviation.

Model 2 also revealed interesting results on mathematics teacher preparation measures. The dichotomous measures indicating whether or not mathematics teachers held an undergraduate degree in a math-related field ($\beta = 1.01$, $p < .001$) and had completed their certification/credential ($\beta = 1.17$, $p < .001$) had moderate and statistically significant effects on mathematics achievement (a combined difference of 20% of a standard deviation). Finally, Model 2 also reveals that when the aforementioned student and teacher characteristics are considered in concert, the effects of whether or not a school is public ($\beta = -0.87$, $p > .1$) or urban ($\beta = -0.76$, $p > .1$) are not statistically significant. However, the effects of two specific features of school contexts, attending a school with a high percentage of low-income students who qualify for free or reduced-price lunch ($\beta = -1.06$, $p < .001$) as well as attending a school with a high percentage of students in the academic track ($\beta = 0.31$, $p < .001$) were statistically significant.

In Figure 7.1, we display the achievement patterns across the four racial-ethnic groups and the subgroups of Asian and Latino LMs. Consistent with our analysis, the positive effects of course taking are displayed in relation to mathematics achievement outcomes on the ELS: 2004 assessment. The

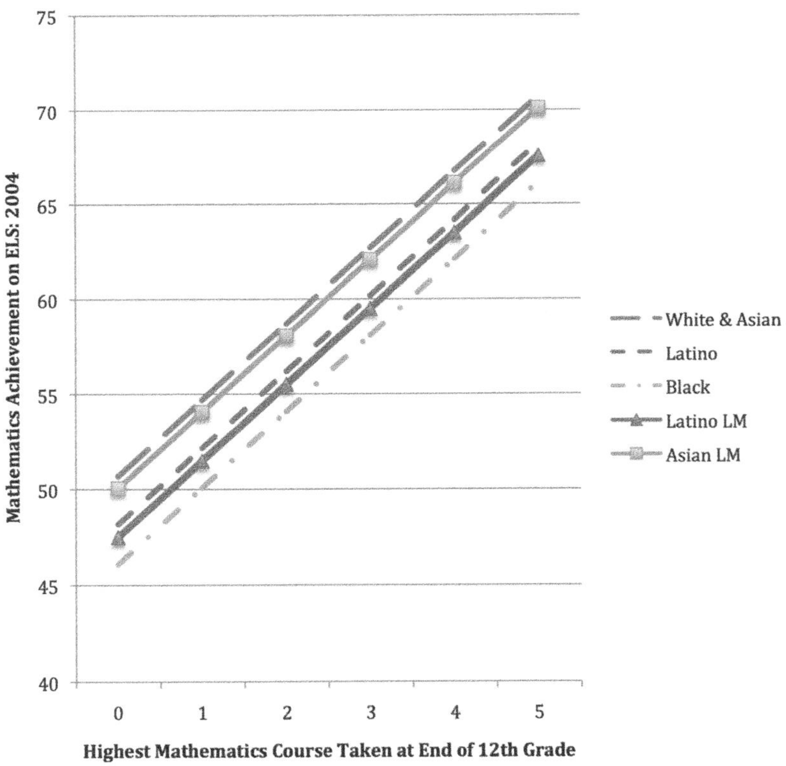

Figure 7.1 Fitted plot displaying ELS: 2004 mathematics achievement differences among Asian, Black, Latino, and White students compared to Asian and Latino LMs relative to their highest mathematics course taken at the end of twelfth grade.

results show that while Asian and White students are the highest achievers, they are followed closely by Asian LMs. The next lowest group is Latinos, followed closely by Latino LMs. Lastly, the fitted plot shows that Black students, on average, are the lowest achievers in this analysis.

Model 3 addresses the second research question, which examines the joint effect of LM status between Asian and Latino students and their ELP. This analysis revealed statistically significant (three-way) interactions between the variables for ethnicity (Latino or Asian), LM status, and ELP in relation to the highest mathematics course taken by the end of twelfth grade. This suggests that the impact of LM status and ELP depends on whether a student is Latino or Asian. Given the nature of interaction terms, one cannot interpret these coefficients alone. Rather, coefficients must be interpreted in conjunction with the main effects of each predictor. The results from Model 3 show that Asian LM students with *low* ELP score slightly below the achievement of Asian LMs with *high* ELP. However, the analysis also shows that the achievement of Latino LMs with low ELP dramatically differs from that of their co-ethnic peers with high ELP, on average. More specifically, the results suggest that Asian LM students scored about one and one-half points below ($\beta = -1.67$, $p < .01$) their White and Asian non-LM peers and, at the same time, their predicted achievement is associated with a positive .015 difference for every unit change in the ELP level, depending on the individual proficiency of LM that ranges from a low value of 2 to a high value of 8. The differences between the tests scores of Latino LMs ($\beta = -5.84$, $p < .05$) relative to White students were not only far lower compared to those of Asian LMs, but the disparities among Latino LMs with low to high levels of ELP were also far more pronounced. For Latino LMs, every one-unit difference in ELP was associated with a .74 difference in mathematics achievement.

Due to the complexity of interpreting statistical interactions, these findings are represented visually in Figure 7.2, holding all control predictors constant at their respective means. The chart in Figure 7.2 builds on Figure 7.1 and adds the interaction effects of LM status for Asian and Latinos by providing a fitted regression trend line for LMs with low (ELP = 3) and high (ELP = 7) levels of ELP. The fitted plot shows that the effect of ELP on the achievement of Asian students with low and high level of ELP is proximal. However, the fitted trend line for the average achievement of Latino LMs with low and high levels of proficiency reflect markedly disparate patterns. While the score of Latino LMs with high levels of ELP are comparable to the scores of Latino non-LMs, the scores of Latino LMs with a low degree of ELP fall below the scores of Black students and are thus the lowest achievers in this study.

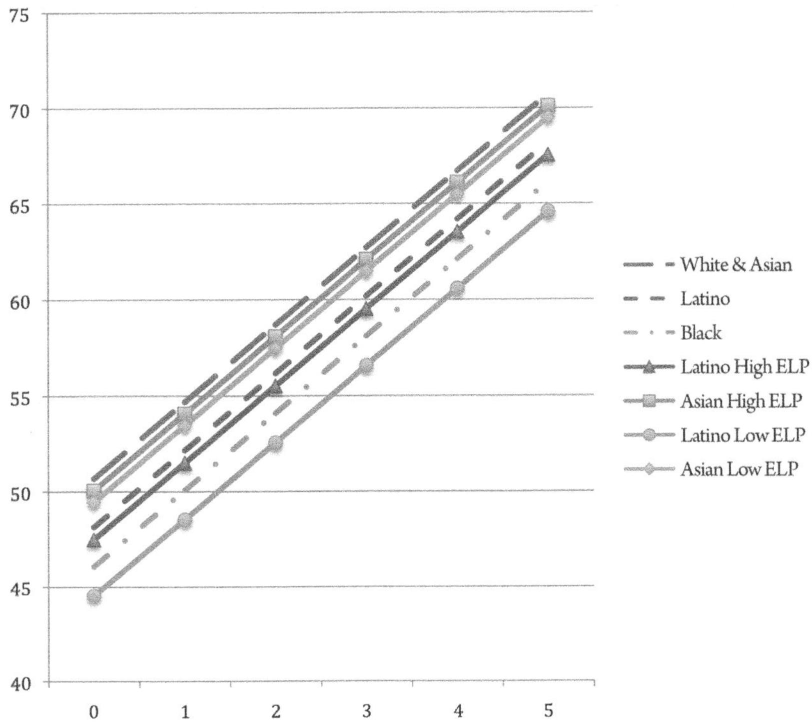

Figure 7.2 Fitted plot displaying ELS: 2004 mathematics achievement main and interaction effects among Asian, Black, Latino, and White students compared to Asian and Latino LMs (with low and high degrees of ELP) relative to their highest mathematics course taken at the end of twelfth grade.

DISCUSSION

The purpose of this chapter was to examine differences in course taking and mathematics achievement among Asian, Black, Latino, and White students and their relationship to urban schools contexts. In addition, we also examined differences across these racial-ethnic groups and between their LM Asian and Latino co-ethnic peers. Our findings suggest that although certain urban school context features had a negative effect on aggregate achievement, all students benefit from access to advanced mathematics courses by the end of their senior year of high school. However, a more nuanced analysis of Asian and Latino LMs illuminates troublesome disparities for Latino LMs with low degrees of ELP.

Consistent with previous research (Orfield & Lee, 2005; Rumberger & Palardy, 2005), our analysis shows that urban school context measures related to poverty concentration within schools negatively impact achievement. In particular, the concentration of low-income high school students was statistically significant and inversely related to mathematics achievement. This suggests that higher proportions of low-income students within a school are correlated with lower aggregate achievement scores, on average. In our analysis of the urban school context, we also controlled for linguistic segregation (as measured by the proportion of ELP students in each school), the proportion of credentialed teachers in each school, and whether the school was located in a rural, suburban, or urban environment. However, these urban school context indicators were not statistically significant once average school SES was introduced into the model.

Interestingly, the proportion of students enrolled in college preparatory courses had a positive effect on mathematics achievement. We found that, on average, schools with higher percentages of students in college preparatory courses were associated with higher aggregate mathematics achievement scores. While providing access to college preparatory courses to larger pools of students can improve mathematics achievement outcomes, the aforementioned literature documented that such opportunities diminish in urban schools (Oakes et al., 2004).

Our results also suggest that the effects of student background characteristics are important indicators of achievement above and beyond the negative effects of attending urban schools. Namely, our findings show that course taking was the most important indicator of mathematics achievement in this study. Each higher mathematics course taken in the college preparatory mathematics sequence was associated with a 4-point higher difference in mathematics achievement. This difference is substantial and totals 40% of a standard deviation.

While a great deal of emphasis is often placed on the English proficiency of LMs as an indicator of low achievement, this study's findings show that having a lower degree of ELP has a negative effect on the mathematics achievement of both Asian and Latino LMs, relative to their non-LM co-ethnics. Nevertheless, Asian LMs outperform Latino LMs, Latino non-LMs, and Black students. This finding is consistent with the aforementioned literature, which has found that LM students with lower levels of English proficiency underperform compared to their English-speaking peers (Callahan, 2005; Mosqueda, 2010; Wang & Goldschmidt, 1999).

However, our analyses (in Model 3) also examined whether the effect of ELP is dependent on whether LM students are Asian or Latino. In other words, does the effect of ELP on mathematics achievement differ by ethnicity? Our results show that for Asian LMs, the effect between high and low ELP is not striking; however, for Latinos, the effect between high and low ELP is dramatically prominent. We find that Latino LMs with low ELP are in fact the most underperforming students in this study. However, as we have discussed above, ELP is but one of many factors that negatively compound the achievement of Latino LMs. In underresourced urban schools, it is imperative that resources are concentrated in areas that foster academic achievement.

CONCLUSIONS

Our findings suggest that in the case of Asian, Black Latino, and White students, limiting access to rigorous coursework can depress their achievement potential in math and potentially curtail their chances for subsequent STEM participation. Because the consequences for missing out on advanced mathematics courses are so great, schools must be vigilant in providing appropriate linguistic supports to maximize achievement in college preparatory courses, particularly in urban schools.

In addition, the common practice of requiring LM students' to gain full English proficiency before they are allowed access to rigorous mathematics content courses must be reconsidered. Rather than delaying opportunities to learn college preparatory mathematics content, LM students require accelerated mathematics courses that integrate explicit scaffolding of the academic language of mathematics in the instructional context of rigorous content. The traditional institutional practice of maintaining LM students in low-level courses ensures that they will remain at the lowest rungs of mathematics achievement.

Urban schools that prioritize addressing students' linguistic needs may potentially improve the mathematics achievement of all students, not just LMs. One of the most referenced reforms in urban school mathematics, Bob Moses' Algebra Project, was developed to primarily address the academic preparation of Black students (Moses & Cobb, 2001). The Algebra Project advocates for a five-step curricular process that includes: 1) A physical event, usually a field trip, 2) pictorial representation/modeling; drawing pictures of features of the physical event, 3) intuitive language; expressing the event in everyday language or "people talk," 4) structured language or "feature talk," isolating specific features of the experience that builds mathematical concepts, and, 5) symbolic representation, which involves conventional mathematical notation (Moses & Cobb, 2001). Two of the five steps of the Algebra Project's curricular process focus explicitly on mathematical language. Consistent with the Algebra Project, our research suggests that language is a very powerful instructional element for all urban school students. Therefore, our work illustrates that taking the most rigorous courses in mathematics must simultaneously involve the explicit teaching of the language of mathematics.

In order to "equalize" the playing field, it is necessary to do more than maximize OTL in urban public schools. In addition to access to riogorous content and qualified teachers, instruction must also be amplified to address the linguistic needs of all students. For all students, the explicit attention to developing the language of the mathematics discipline is critical. For LMs, and those with low degrees of ELP in particular, providing English-language development within mathematics courses is essential to making rigorous content comprehensible.

The unquestioned factors attributed to enduring achievement disparities of students in urban schools all too often result in state takeovers and imminent school closures. Our analysis identifies several factors that influence such disparities: ELP and access to both rigorous college preparatory courses and qualified teachers. However, this study reconceptualizes such factors as simultaneous achievement determinants and as indicators of the types of supports, resources and services necessary to raise mathematics achievement in urban school contexts. In other words, as long as the neglect of these critical factors continues, enduring achievement disparities will remain steadfast.

This research has important implications for improving mathematics achievement in urban schools, a discipline that is foundational to post-secondary opportunities in STEM. First and foremost, mathematics instruction in urban schools needs to accelerate access to rigorous college-preparatory courses while concurrently fostering the language of mathematics. This study confirms the undeniable relationship between achievement and rigorous coursework; the highest achieving students were those who completed the most rigorous courses by the end of high school. Consistent with our findings, the positive effect of larger proportions of students enrolled in college preparatory mathematics attenuated the negative effect of poverty concentration in urban schools. Secondly, although both Asian and Latino LMs scored lower than their non-LM coethnics, the negative effect of language on achievement was more detrimental for Latinos than for Asians. This corroborates the claim that urban schools serving large proportions of Latino LMs must make mathematics content and the language of mathematics comprehenshible through English-language development strategies, without compromising access to rigorous content. Finally, in order to reverse inequitable achievement trends in urban schools, resources must be invested towards building the capacity of mathematics teachers through professional development that integrates rigorous content, the language of the mathematics discipline, and English-language development for LMs. Endeavoring to improve achievement in urban schools requires unwavering commitment to teacher professional development to ensure equity and access in STEM education.

NOTES

1. They include the unavailability of qualified teachers, inadequate instructional materials, and overcrowded facilities.
2. Widely cited large-scale sociological studies of immigrants using similar types of datasets have used these same self-reported English proficiency measures and find that they are relatively reliable measures of language skills (Portes & Rumbaut, 2001).
3. The principal components analysis routine in STATA yielded the following weighted composite equation: ENG_PROF = .486 × UNDERSTAND + .511 × SPEAK + .510 × READ + .492 × WRITE. This single construct of English proficiency captured 68% of the variance in the four English proficiency subscales.
4. The questionnaire weight (for F1) applies to all first follow-up respondents.
5. The intraclass correlation (ρ) partitions the variance in the outcome due to between-institution (τ_{00}) and within-institution (σ^2) differences, and is calculated by the following formula: $\rho = \tau_{00}/(\tau_{00} + \sigma^2)$.
6. We calculated the partitioning of variance between Level 1 and Level 2 by $\rho = 23.15/(23.15 + 54.6)$.
7. γ_1 is a parameter vector describing the impact of the individual-level controls Z_{ij}.
8. γ_2 is a parameter vector describing the impact of the school-level controls Z_j.

REFERENCES

Abedi, J. (2004). The No Child Left Behind Act and English language learners: Assessment and accountability issues. *American Educational Research Association, 33,* 4–14.

Abedi, J., & Gándara, P. (2006). Performance of English language learners as a subgroup in large-scale assessment: Interaction of research and policy. *Educational Measurement: Issues and Practice, 25,* 36–46.

Adelman, C. (1999). *Answers in the tool box: Academic intensity, attendance patterns, and bachelor's degree attainment.* Washington, DC: U.S. Department of Education.

Adelman, C. (2006). *The toolbox revisited: Paths to degree completion from high school through college.* Washington, DC: U.S. Department of Education.

American Educational Research Association. (1999). *Standards for educational and psychological testing.* Washington, DC: Author.

August, D., & Hakuta, K. (1997). *Improving schooling for language-minority children: A research agenda.* Washington, DC: National Research Council, Institute of Medicine, National Academy Press.

Boaler, J. (2008). *What's math got to do with it?: Helping children love their most hated subject and why it's important for America.* London, UK: Penguin.

Callahan, R. (2005). Tracking and high school English learners: Limiting opportunity to learn. *American Educational Research Journal, 42,* 305–328.

Callahan, R. M., Wilkinson, L., & Muller, C. (2010). Academic achievement and course taking among language minority youth in U.S. schools: Effects of ESL placement. *Educational Evaluation and Policy Analysis, 32*(1), 84–117.

Callahan, R. M., Wilkinson, L., Muller, C., & Frisco, M. (2009). ESL placement and schools: Effects on immigrant achievement. *Educational Policy, 23*(2), 355–384.

Capraro, M. M., Capraro, R. M., Yetkiner, Z. E., Rangel-Chavez, A. F., & Lewis, C. W. (2010) Examining Hispanic students mathematics performance on high-stakes tests: An examination of one urban school district in Colorado. *The Urban Review, 42*(3), 193–209.

Capraro, R. M, Young, J. R., Lewis, C. W, Yetkiner, Z. E., & Woods, M. N. (2009). An examination of mathematics achievement and growth in a midwestern urban school district: Implications for teachers and administrators. *Journal of Urban Mathematics Education, 2*(2), 46–65.

Cuevas, G. (1984). Mathematics learning in English as a second language. *Journal for Research in Mathematics Education, 15,* 134–144.

Darling-Hammond, L., Berry, B., & Thoreson, A. (2001). Does teacher certification matter? Evaluating the evidence. *Educational Evaluation and Policy Analysis, 23*(1), 57–77.

Darling-Hammond, L., & Youngs, P. (2002). Defining "highly qualified teachers": What does "scientifically-based research" actually tell us? *Educational Researcher, 31*(9), 13–25.

Dentler, R. A., & Hafner, A. L. (1997). *Hosting newcomers: Structuring educational opportunities for immigrant children.* New York, NY: Teachers College Press.

Flores, A. (2007). Examining disparities in mathematics education: Achievement gap or opportunity gap? *The High School Journal, 91*(1), 29–42.

Frankenberg E. (2009). The demographic context of urban schools and districts. *Equity and Excellence in Education, 42*(3), 255–271.

Gándara, P., & Contreras, F. (2009). *The Latino education crisis: The consequences of failed social policies.* Cambridge, MA: Harvard University Press.

Gándara, P., & Maxwell-Jolly, J. (2006). Critical issues in developing a teacher corps for English learners. In K. Téllez & H. C. Waxman (Eds.), *Preparing quality educators for English language learners: Research, policies, and practices* (pp. 99–120). Mahwah, NJ: Erlbaum.

Gándara, P., & Orfield, G. (2010). A return to the "Mexican room": The segregation of Arizona's English learners. *The Civil Rights Project/*Proyecto Derechos Civiles at UCLA. Los Angeles, CA. Retrieved from http://civilrightsproject.ucla.edu/research/k-12-education/language-minority-students/a-return-to-the-mexican-room-the-segregation-of-arizonas-english-learners-1

Garrison, L. & Mora, J. K. (1999). Adapting mathematics instruction for English language learners: The language concept. In L. Ortiz-Franco, N. G. Hernandez, & Y. De La Cruz (Eds.), *Changing the faces of mathematics: Perspectives on Latinos.* Reston, VA: National Council of Teachers of Mathematics.

Gifford, B., & Valdés, G. (2006). The linguistic isolation of Hispanic students in California's public schools: The challenge of reintegration. *Yearbook of the National Society for the Study of Education, 105*(2), 125–154.

Gutiérrez, R. (2002). Beyond essentialism: The complexity of language in teaching mathematics to Latina/o students. *American Educational Research Journal, 39,* 1047–1088.

Gutiérrez, R. (2007). (Re)defining equity: The importance of a critical perspective. In N. S. Nasir & P. Cobb (Eds.), *Improving access to mathematics: Diversity and equity in the classroom* (pp. 37–50). New York, NY: Teachers College Press.

Hanushek, E. A. (1992). The trade-off between child quantity and quality. *Journal of Political Economy, 100*(1), 84–117.

Ingels, S. J., Pratt, D. J., Rogers, J. E., Siegel, P. H., & Stutts, E. S. (2004). *Education longitudinal study of 2002: Base year data file user's manual.* Washington, DC: U.S. Department of Education, National Center for Education Statistics.

Khisty, L. (1995). Making inequality: Issues of language and meanings in mathematics teaching with Hispanic students. In W. G. Secada, E. Fennema, & L. B. Adajian (Eds.), *New directions for equity in mathematics education* (pp. 279–298). New York, NY: Cambridge University Press.

Kozol, J. (1991). *Savage inequalities: Children in America's schools.* New York, NY: Crown.

Kretovics, J., & Nussel, E. J. (1994). *Transforming urban education.* Needham Heights, MA: Allyn & Bacon.

Lee, V. E., & Bryk, A. S. (1988). Curriculum tracking as mediating the social distribution of high school achievement. *Sociology of Education, 61*(2), 78–94.

Lee, V. E., Croninger, R. G., & Smith, J. B. (1997). Course-taking, equity, and mathematics learning: Testing the constrained curriculum hypothesis in U.S. secondary schools. *Educational Evaluation and Policy Analysis, 19,* 99–121.

Light, R., Singer, J. D., & Willett, J. B. (1990). *By design: Planning research in higher education.* Cambridge, MA: Harvard University Press.

Lucas, T., Henze, R., & Donato, R. (1990). Promoting the success of Latino language-minority students: An exploratory study of six high schools. *Harvard Educational Review, 60,* 315–340.

Maeroff, G. I. (1988, May). Withered hopes, stillborn dreams: The dismal panorama of urban schools. *Phi Delta Kappan, 69*(9), 633–638.

Maxwell-Jolly, J., & Gándara, P. (2002). A quest for quality: Providing qualified teachers for California's English learners. In Z. F. Beykont (Ed.), *The power of culture: Teaching across language difference.* Cambridge, MA: Harvard Education Publishing Group.

Minicucci, C., & Olsen, L. (1992). *Programs for secondary limited English proficient students: A California study.* Washington, DC: National Clearinghouse for Bilingual Education.

Moses, R. P., & Cobb, C. E., Jr. (2001). *Radical equations: Math literacy and civil rights.* Boston: Beacon Press.

Mosqueda, E. (2010). Compounding inequalities: English proficiency and tracking and their relation to mathematics performance among Latina/o secondary school youth. *Journal of Urban Mathematics Education, 3*(1), 57–81.

National Center for Education Statistics (NCES). (2011, May). *The condition of education 2010. Indicator 6: Children who spoke a language other than English at home.* Washington, DC: U.S. Department of Education. Retrieved from http://nces.ed.gov/pubs2011/2011033.pdf

Noguera, P. (1996). Confronting the urban in urban school reform. *The Urban Review, 28*(1), 1–19.

Noguera, P. A. (2003). *City schools and the American dream.* New York, NY: Teachers College Press.

Oakes, J. (1985). *Keeping track: How schools structure inequality.* New York, NY: Yale University Press.

Oakes, J., Rogers, J., Silver, D., Horng, E., & Goode, J. (2004). *Separate and unequal. 50 years after Brown: California's racial opportunity gap.* Los Angeles, CA: Institute for Democracy, Education, and Access, University of California-Los Angeles.

Olsen, L. (1995). School restructuring and the needs of immigrant students. In R. G. Rumbaut & W. A. Cornelius (Eds.), *California's immigrant children: Theory, research, and implications for educational policy* (pp. 209–232). San Diego, CA: Center for U.S.–Mexican Studies, University of California.

Olsen, L. (1997). *Made in America: Immigrant students in our public schools.* New York, NY: New Press.

Orfield, G., & Lee, C. (2005). *Why segregation matters: Poverty and educational inequality.* Cambridge, MA: Harvard University, The Civil Rights Project.

Pelavin, S. H., & Kane, M. (1990). *Changing the odds: Factors increasing access to college.* New York, NY: College Board.

Portes, A., & Rumbaut, R. G. (2001). *Legacies: The story of the immigrant second generation.* Berkeley, CA: University Press and Russell Sage.

Raudenbush, S. W., & Bryk, A. S. (2002). *Hierarchical linear models: Applications and data analysis methods* (2nd ed.). Newbury Park, CA: Sage.

Ruiz-de-Velasco, J., & Fix, M. (2000). *Overlooked & underserved: Immigrant students in U.S. secondary schools.* Washington, DC: Urban Institute.

Rumberger, R. W., & Palardy, G. J. (2005). Does segregation still matter? The impact of social composition on academic achievement in high school. *Teachers College Record, 107*(9), 1999–2045.

Secada, W. G. (1992). Race, ethnicity, social class, language, and achievement in mathematics. In D. A. Grouws (Ed.), *Handbook of research on mathematics teaching and learning* (pp. 623–660). New York, NY: Macmillan.

Secada, W. G. (1996). Urban students acquiring English and learning mathematics in the context of reform. *Urban Education, 30*(4), 422–448.

Smith, T. M. (1995). *The educational progress of Hispanic students* (NCES 95-767). Washington, DC: U.S. Department of Education, Office of Educational Research and Improvement.

Solano-Flores, G. (2011). Language issues in mathematics and the assessment of English language learners. In K. Téllez, J. N. Moschkovich, & M. Civil (Eds.), *Latinos and mathematics education: Research on learning and teaching in classrooms and communities.* Charlotte, NC: Information Age.

Tate, W. F. (1997). Race-ethnicity, SES, gender, and language proficiency trends in mathematics achievement: An update. *Journal for Research in Mathematics Education, 28,* 652–679.

Téllez, K., & Waxman, H. C. (2006). *Preparing quality educators for English language learners: Research, policies, and practices.* Mahwah, NJ: Erlbaum.

Valdés, G. (1998). The world outside and inside schools: Language and immigrant children. *Educational Researcher, 27*(6), 4–18.

Valencia, R. R. (2002). *Chicano school failure and successes: Past, present, and future.* London, UK: Rutledge.

Valenzuela, A. (1999). *Subtractive schooling: U.S.-Mexican youth and the politics of caring.* Albany, NY: State University of New York Press.

Wang, J., & Goldschmidt, P. (1999). Opportunity to learn, language proficiency, and immigrant status effects on mathematics achievement. *Journal of Educational Research, 93*(2), 101–111.

Wilkins, A., & Education Trust. (2006). *Yes we can: Telling truths and dispelling myths about race and education in America.* Washington, DC: Education Trust.

Wilson, W. J. (1987). *The truly disadvantaged: The inner city, the underclass, and public policy.* Chicago, IL: Chicago University Press.

Wong-Fillmore, L. (2007). English learners and mathematics learning: Language issues to consider. In A. H. Schoenfeld (Ed.), *Assessing mathematical proficiency* (pp. 333–344). Cambridge, MA: Cambridge University Press.

CHAPTER 8

USING THREE-DIMENSIONAL VIRTUAL ENVIRONMENTS TO PREPARE STEM TEACHERS

Trina Davis
Texas A & M University

There are prominent shortages that the United States faces related to the preparation and later recruitment of K–12 students into science, technology, engineering, and mathematics (STEM) postsecondary programs. Equally alarming are the disparities in recruiting underserved groups (e.g., women and minorities) into STEM fields (Bayer Corporation, 2010). In addition, the gaps in achievement across STEM subject areas between White students and students of color are an ongoing challenge for educators (Flores, 2007). The combination of deep and specific content knowledge and a thorough understanding of diverse learners is necessary for preparing teachers to close the achievement gaps in STEM (Brown, Davis, & Kulm, 2011). Exploring new and innovative ways to prepare teachers to engage *all* students in rich and effective learning experiences in STEM offers promise in addressing these parallel needs. To this end, teachers must be well prepared to design and deliver effective STEM instruction that meets the needs of all students.

Improving Urban Schools, pages 125–141

An important component of teacher preparation programs are to provide knowledge and experience that provide a realistic preparation for the classroom. Moreno and Ortegano-Layne (2008) note that we learn from directly observing (real) people as well as by indirectly observing people in real or fictitious situations, like motion pictures, television, plays, or books. They further report that preservice teachers have chosen classroom scenarios as their preferred method of instruction over more traditional teacher education methods (Moreno & Ortegano-Layne, 2008).

> According to experiential models of learning, when students are presented with a classroom scenario demonstrating how the learned principles can be applied to the teaching practice, relevant aspects of the scenario are selected by matching the encoded principles with observed/described classroom behaviors and the example is integrated with students' past experiential knowledge. (p. 452)

In recent years, researchers and practitioners have begun to explore innovative ways to prepare teachers through the use of virtual learning environments (VLEs). For example, Brown et al.'s (2011) project focuses on designing and testing virtual learning strategies that enhance preservice teachers' mathematical knowledge for teaching algebra in the middle grades. The design and use of a virtual classroom in *Second Life* (SL) are at the center of their *Knowledge for Algebra Teaching for Equity (KATE) Project* (2012). Project team members designed simulations of classroom situations that challenged preservice teachers to address equity issues in their design and delivery of microlessons in a virtual classroom setting. Researchers assert that these simulations can provide more directed practice with diverse learners than the periodic opportunities that might occur in classroom observations or other field-based settings. Ultimately, programs aimed at providing early teaching experiences in simulated classroom environments show great promise in preparing teachers.

Background on Simulations and Virtual Worlds

Three-dimensional (3-D) virtual learning environments, such as the virtual world of *Second Life*, are available, which can support STEM learning and teacher preparation in new and exciting ways. Many concepts and competencies that are foundational across various STEM fields require unique understandings. Computational thinking and problem solving, for example, are critically important across various fields, with an emphasis in technology, engineering, and mathematics certainly. Similarly, various concepts across the field of science require the understanding of spatial relationships (Merchant et al., 2012). One can find engaging simulations in *Second Life* that provide powerful models for STEM teaching and learning. The *KATE* project enriches the education of STEM teachers by using emergent technologies like *Second Life* to provide preservice middle school mathematics teachers early teaching experiences that address topics in problem solving, algebra, and educational equity. Also, *The Molecule Game,* found on *Dr. K's Chemistry Corner* in *Second Life,* which was designed for students to explore molecules in a 3-D space, provides another promising example. These illustrative classroom simulation examples will be discussed in greater detail later in the chapter. First, background and perspectives will be presented related to (a) simulations and role-playing, (b) defining virtual worlds and their affordances, (c) virtual worlds in teaching and learning, and (d) then specifically virtual worlds in teaching and learning STEM. These perspectives will be illuminated and connections will be made from a teacher-preparation context.

Simulations and Role-Playing

Simulations as a subset of computer-assisted learning have been around since the late 1980s (e.g., SimCity) (Wright, 1989). These earlier programs typically simulated real-life places and scenarios. They provided opportunities for learners to interact in environments even when they were not able to physically visit them. In this way, 3-D virtual environments were an effective alternative and fostered experiential and situated learning approaches (Davis, 2012b). Virtual worlds such as *Second Life* are on the leading edge of a new set of technological and experiential applications that have the potential to impact how people communicate, play, work, and, most relevantly, learn. Advances in educational theory and cognitive science mean more is understood about the process and impact of learning than ever before. Much of the work on learning in virtual worlds has been exploratory. There are many exciting possibilities for using *Second Life* as a learning tool (Goodband, Bhakta, & Lawson, n.d.). It is critically important that preservice teachers have hands-on opportunities to experience learning and teaching in these emergent virtual environments. In addition to providing rich classroom teaching experiences for preservice teachers, it is only through these early experiences in their preparation programs that they will gain comfort, fluency, and insight in the potential of using these resources to engage their students in new and immersive ways, as well.

While being actively engaged in experiential and situated learning in virtual environments, learners can be afforded ubiquitous opportunities to authentically develop an array of skills such as (a) new literacies (Jenkins, Clinton, Purushotma, Robinson, & Weigel, 2006), (b) problem-solving or mathematics skills (National Council of Teachers of Mathematics, 2000),

(c) scientific literacy skills, and (d) computational thinking as well as information and communication technology (ICT) skills (Barr, Harrison, & Conery, 2011). A sample of these skills includes but is not limited to collaboration, problem solving, simulation, critical thinking, and negotiation. The new literacies, for example, almost all involve social skills developed through collaboration and networking (Jenkins et al., 2006).

According to Caprotti and Seppala (2007), the design of 3-D learning activities in *Second Life* is not hard, although activities that focus on modeling and working out mathematics problems may require different approaches or instructional planning. For example, with recent developments in SL, such as extended use of displays that project an array of formats including audio, video, and web-based media, and now the availability of interactive pen displays (i.e., *Smart Podium* solutions) paired with streaming applications, mathematics concepts, symbolic representations, and problems, can be presented or solved with greater ease (Davis, 2012b). These recent developments improve the utility for preservice teachers to be able to work out problems as they give lessons in virtual classrooms in *Second Life*.

Slightly different from simulation, role-playing requires more specific concepts on interactive points of view for enhancing interpersonal relations and social transaction among individuals (Tompkins, 1998). Role-playing, as Scarcella and Oxford (1992) defined it, is acting out a character represented somehow by everyday-life experiences. To achieve the goals for the target topics, participants need to follow the intended *mission* and responsibilities in order to immerse themselves in situations directed toward the ultimate goals (Jones, 1982). With the burgeoning development of online education, role-playing in online contexts has increased in its popularity (Gao, Noh, & Koehler, 2008). Beginning with text-based online role-playing, a number of empirical studies indicate the benefits of using role-play for developing comprehensive understandings and unpacking complex concepts (Bell, 2001; Freeman & Capper, 1999; Gao et al., 2008). The affordances of employing role-playing in *Second Life* can facilitate the design of powerful, more authentic classroom experiences for preservice teachers. Highlights from the *KATE* project illustrate this application nicely later in the chapter.

Defining Virtual Worlds

A virtual world (VW) is an online simulation and computer-based program wherein users can interact with other users, objects, content, and materials via virtual surrogates. In VWs, real-world objects and imaginative materials are graphically and multimedia rich. With their surrogates (or avatars), users can move, build, create, and modify content and material freely. Through the Internet, the VW is usually presented as a three-dimensional multiuser environment where online synchronous and asynchronous communications become possible. Virtual worlds are participatory environments that can simulate physical presence in places in the real world, as well as in imaginary settings.

A number of definitions on virtual worlds contribute to a more complete picture. Castronova (2004) describes VWs as synthetic worlds stored in online networking for accommodating a large number of people. This description represents only minor parts of features in VWs. Bartle (2004) explains that with Internet accessibility, VWs can be shared in multiuser spaces. He also notes that they contain a variety of real-world and imaginary content. Bell's (2008) definition is most comprehensive. He defines a virtual world as "a synchronous persistent network of people represented as avatars, and facilitated by a network of computers" (p. 2). He explains that the synchrony of VWs means the online persistency continues to process without being able to be stopped and paused as long as the Internet connection is accessible. Virtual worlds have extensive social networking components as well and facilitate the connection of users to myriad broader social and professional communities.

Virtual Worlds in Teaching and Learning

In recent years, increasing numbers of educators and stakeholders have seen the promise of harnessing the affordances of VWs in teaching and learning. Wang and Woo (2007) suggest that a downside of face-to-face classes is students' anxiety and lack of ease in accessing course materials. Thompson (2008) suggests that a concern with some web-based classes is the lack of communication and socialization. Some maintain that VWs are capable of addressing some of the disadvantages of traditional face-to-face or web-based classes (Dickey, 2005; Dillenbourg, Schneider, & Synteta, 2002; Kluge & Riley, 2008). Caprotti and Seppala (2007) argue that VWs have the potential to address some of the social issues associated with online learning: "Students studying in online courses using their computers lack social interaction with their fellow students. If they attend virtual classes as avatars, they can, at least, see other avatars, and talk to them between classes" (p. 1).

VWs provide rich visual materials, content, and objects in a 3-D environment (Dickey, 2005). The graphically intensive nature of the environment facilitates different ways of teaching and learning, especially for visual learners (Annetta & Park, 2006). Virtual worlds also allow users to freely customize a variety of virtual objects for visual learners to better understand topics (McComas, Pivik, & Laflamme, 1998). Barab, Thomas, Dodge, Carteaux, and Tuzun (2005) argue that VWs provide “rapid visual prototyping” for educators to understand and respond to learners’ preferences. Moreover, Minocha and Reeves (2010) discovered that the visual realism of a learning space can have impact on a learner’s motivation through visually rich learning experiences and activities.

Although the utilization of virtual worlds in higher education is widespread, the empirical research on preservice teacher training in SL is a new territory yet to be investigated thoroughly (Cheng, Zhan, & Tsai, 2010). In a study conducted by Gregory and Masters (2010), 90 preservice teachers (PSTs) without any experience using SL were introduced to using SL to simulate real-life environments. Although the results revealed that PSTs enjoyed the real-life workshops more than the SL workshops and found the SL experiences more difficult, the study found a positive perception of using SL as an effective teacher training tool.

In a qualitative study using SL for PST training for language teaching (Jauregi, Canto, de Graaff, Koenraad, & Moonen, 2011), Spanish PSTs indicated that communication in the virtual environment and topic-specific settings in SL are advantages they enjoyed more than video conferencing. Moreover, in a cooperative study in which 65 Mandarin PSTs in a university in Taiwan taught 13 American learners of Mandarin in an SL setting, the results showed that the majority of PSTs enjoyed using avatars and believed that SL offers an ideal setting for language communication, teaching, and learning. The study concluded that SL provides a feasible and meaningful context for Mandarin PSTs in Taiwan to teach learners of Chinese in the United States (Cheng et al., 2010). In their study simulating classroom misbehaviors, Mahon, Bryant, Brown, and Kim (2010) describe advanced utilizations of SL by applying artificial intelligence (AI). This study immersed PSTs in a simulated classroom in which 30 avatars were controlled by game-based AI scripting that generated random misbehaviors. Researchers integrated diverse cultural and linguistic backgrounds to AI avatars, and PSTs had opportunities to practice how to react and respond to different misbehaviors. The study found SL to be an effective simulation of a real-life classroom setting in which PSTs practiced dealing with classroom management and behavior issues.

Virtual Worlds in STEM Teaching and Learning

There are a number of affordances that VWs offer that can lend themselves to STEM teaching and learning (Dalgarno & Lee, 2010). Although there are growing numbers of higher education simulations and projects in *Second Life*, there is still room for greater penetration in STEM areas. If one explicitly conducts a search on STEM in *Second Life*, it will yield two prominent results: *STEM Island* (which houses a virtual STEM exhibition) or the *Marshall University Virtual Campus and STEM Academy*. Searching across STEM areas will yield additional results. Caprotti and Seppala (2007) report that searching for mathematics in SL yields very few locations or projects that focus solely on mathematics. However, as one looks to the literature, there are examples of empirical work on teaching and learning mathematics in virtual worlds. Some studies suggest that mathematics exploration in VWs not only provides students visual information to see the mathematics logic (e.g., 3-D geometric construction) (Kaufmann & Schmalstieg, 2003), but also have the potential to increase students' engagement and effectiveness in learning mathematics (Harrell et al., 2008).

Educational Virtual Environment-Mathematics (EViE-M) provides an example of designing a virtual reality platform for mathematics (Malamos et al., 2008). It is designed for learners to acquire mathematical experience through problem solving. EViE-M is a virtual environment where users construct an organized city through answering math questions for gaining access to buildings. In another project, Murfin (2001) describes a case study that is situated in a mathematics class using microcomputers and computer animation. The collaborative virtual learning environment is called Science Multi-User Dimension Object Oriented (ScienceMOO). The majority of participating students expressed positive attitudes toward the place-independent nature of ScienceMOO and increased communication and interaction. Some students expressed negative attitudes toward technical problems, text overload, and lack of familiarity.

STEM Project and Simulation Examples in Virtual Worlds

The affordances of using 3-D virtual environments to foster rich instructional settings have been discussed throughout the chapter. Features of immersive 3-D virtual worlds can be influential in potentially impacting various learning outcomes across STEM areas. Promising VW projects and simulations across STEM subjects that engage learners in new and different ways are highlighted below. These projects help to illustrate a couple of approaches using VWs to prepare STEM teachers and using VWs to teach STEM content in participatory and immersive ways.

The *KATE* project focuses on equity in algebra learning. Algebra has emerged as a "filter" that prevents many students from pursuing advanced mathematics courses in high school and therefore limits their preparation for college and/or selection of STEM careers. In this 5-year project, the research team continues to develop and test prototypes of virtual learning strategies designed to enhance preservice teachers' mathematical knowledge for teaching emergent concepts of algebra in the middle grades. The project focuses on knowledge of three primary early algebra concepts: change, variables, and operations (American Association for the Advancement of Science, 2000; Kulm & Capraro, 2008).

During the first phase of the *KATE* project, a virtual middle school classroom and additional learning spaces were developed in *Second Life* (see Figure 8.1). Graduate research assistants developed avatars of diverse middle-grades students to simulate their mathematics knowledge and misconceptions in an authentic (virtual) classroom setting. Preservice teachers also developed personal (teacher) avatars. The project is situated within a semester-long *Problem Solving in Mathematics* course at a large Southwestern university with very high research activity (RU/VH) (Carnegie Foundation, 2010). The *Problem Solving in Mathematics* course is a required course taken by all preservice teachers seeking mathematics certification.

Multiple *Second Life*–embedded training and initial classroom simulation sessions were integrated in the *Problem Solving in Mathematics* course throughout the semester prior to student delivering their *SL Problem-Solving Lessons* to the full class. These included the *Intro to Second Life Orientation, Meet Your Middle Grade Students* exercise, and *Tutoring Problem-Solving in Second Life*, and *Demonstration of SL Teaching Example Video* exercises. During the *Meet Your Middle Grade Students* exercise (Figure 8.2), for example, each graduate research assistant (role-playing a middle-grade student [MGS]), met one-on-one with the preservice teachers (PSTs) assigned to them. The MGSs brought their school records. PSTs came with questions they wanted to ask the MGSs about their interests, background, math experiences. The PSTs would later use this information and experience to help plan their

Figure 8.1 *KATE* virtual classroom during a student SL problem-solving lesson.

Figure 8.2 *KATE* middle school classroom during meet-the-student exercise.

one-on-one *Tutoring Problem-Solving in Second Life* lesson as well as their culminating *SL Problem-Solving Lesson* that would be delivered to the full class.

These simulations make it possible to provide more intensive and focused practice with diverse learners (i.e., diverse middle school student actors) than the more sporadic opportunities that might arise in classroom observations or other traditional field-based settings. For example, graduate students serving as MGSs were able to defuse a variety key math misconceptions throughout each of the lessons that were designed and delivered by preservice teachers (Figure 8.3). The approaches employed in the *KATE* project have great potential for generalization to other topic areas, specifi-

Figure 8.3 Preservice teacher giving SL problem-solving lesson.

cally in mathematics or across various STEM topics, and to other teacher education programs.

Dr. K's Chemistry Corner provides an example of how *Second Life* is being used in a core introductory chemistry course. Prominent simulations in this space include the *Molecule Game* and the *Chemist as Artist. The Molecule Game* was designed for students to explore molecules in a 3-D space (Figure 8.4). They can rotate the molecule to view it from different perspectives. This virtual space has been designed to allow the learner to manipulate the molecules and link or unlink the atoms to explore the molecules in detail.

Merchant et al. (2012) describe the *Molecule Game* simulation as allowing students the opportunity to view molecules with their bond angles from various perspectives. They can use the zoom-in and zoom-out features in *Second Life* for multiple views of the molecules. They can also readily examine bond relationships between atoms within a molecule using different capabilities within the SL virtual environment; this includes rotating and manipulating various molecules (Merchant et al., 2012).

Similarly, the *Chemist as Artist* simulation was designed to further develop students' ability to see molecules from various vantage points in an immersive environment. Learners (avatars) are able to photograph themselves next to different orientations of the molecules (using the SL Snapshot feature). They also construct 2-dimensional drawings (representations) of the various orientations.

There is some evidence that the kinds of *Second Life* uses (interventions) that were employed in both the *KATE* project and *Dr. K's Chemistry Corner* offer promising results with preservice teachers or students enrolled in the introductory college chemistry class. Merchant et al. (2012) conducted a study with chemistry students who were engaged in various activities at *Dr.*

Figure 8.4 Dr. K's Chemistry Corner molecule game.

Figure 8.5 Dr. K's Chemistry Corner.

K's Chemistry Corner (Figure 8.5). They examined a model of the impact of using a 3-D desktop virtual environment on learner characteristics (i.e., perceptual and psychological variables), which can enhance chemistry-related learning achievements. The results indicated that 3-D virtual reality–based instruction is effective for enhancing students' chemistry achievement.

Davis (2012) investigated experiences and perceptions of *KATE* project preservice teachers and graduate students serving as middle-grade student actors (avatars), following simulation and engagement exercises in a virtual classroom setting in *Second Life.* Multiple SL classroom simulation teaching sessions were integrated (as described earlier). Generally, over 90% of all preservice teachers enrolled in the required mathematics certification course found the various exercises helpful in allowing them to develop their teaching skills, and they also found the exercises useful in helping them to further develop SL skills. PST's responses to open-ended questions provide several insights on the benefits of using SL to prepare teachers (see Table 8.1).

CONCLUDING THOUGHTS

The value of using *Second Life* to broaden preservice teachers' experiences in innovative teaching and learning settings with diverse students, in some cases, has only begun to be explored. The SL classroom and learning simulations highlighted here seem promising in their ability to add to the richness of instruction, engagement, and overall preparation for preservice teachers. The implementation of the work highlighted suggests that virtual worlds can provide rich settings for teacher preparation and development in STEM areas. Since STEM teacher-preparation projects are just begin-

TABLE 8.1 Preservice Teachers' Perceived Benefits from SL Experiences (Davis, 2012a)

Preservice Teachers' Perceived Benefits
PSTs felt SL engagement and teaching experiences in the course provided them with critically needed first teaching opportunities. PSTs often reported that they felt virtual teaching experiences helped them to see the challenges of teaching MGSs with diverse needs and various math misconceptions. They also felt they had more freedom to explore (and not do any damage) since they were teaching in a simulated environment. Some commented that they liked that real students weren't on the other end.
"I really liked being able to use the smart board. This way the students could see what I was trying to explain verbally. At the end of the presentation, the students began to ask questions, and it was a good experience to see what types of questions middle-grade students may ask. It was an experience that I feel helped me prepare to be in a real classroom."
"I thought it was helpful because it gave us an opportunity to actually use Second Life and figure out how to work with it. It also gave me an opportunity to meet the students so that I could use their names and interests in future problems." (Get to Know Your Student Exercise)
"The final project was helpful. I liked that we did it in a different place where not everyone was watching us. This made it not as stressful and felt like we were really teaching middle-grade students and not college students. I ended up enjoying this lesson a lot more than I originally thought I would."
"It went better than I thought it would because I felt like I responded to students in an effective way and took their suggestions and questions into consideration in my delivery rather than disregarding what they were doing in order to stay exactly on track with the plan that I had. Overall, it helped me to better understand how to assess how students are doing and what they need help with before trying to stick with the plan that I have going into the lesson."
"This was a very cool experience, and I was really able to get the feel of how it would be to teach in a classroom, which I had not experienced before."
"Second Life is great for a couple different reasons. It gives you an opportunity to just see students as students, instead of like what they're wearing or what they look like or anything like that. So it's great in that aspect, and also for, like, we've talked about, they were grad students. So if we did it wrong, it's not like their perceptions of math were totally just messed up forever."

ning to emerge in SL, it appears that this area is fertile ground for trying out new and different approaches. Practitioners and researchers can also draw from models and approaches used in earlier projects as a jumping off point, like those highlighted here. The following section provides recommendations for educators interested in exploring the use of VWs in general or to prepare STEM teachers. Recommendations are included for participants who are new to VWs and for those who are more experienced.

RECOMMENDATIONS

One of the benefits of using a virtual world like *Second Life* is the unlimited access one has to networks of users with similar interests. For educators interested in learning more, there are getting-started steps that can be taken to explore the use of *Second Life* in various educational contexts, or specifically to prepare STEM teachers. Dawley (2009) presents a matrix for helping guide instructors on the introduction and use of social networks in SL. Dawley's social network knowledge construction (SNKC) engagement levels include *identify networks, lurk, contribute, create,* and *lead.* Table 8.2 captures this work and includes additional recommendations for both new and experienced users.

TABLE 8.2 Recommendations for Instructors as Well as New and Experienced Educators Using Virtual Worlds (Dawley, 2009)

Social network engagement level	Recommendations for new users to virtual worlds	Recommendations for more experienced virtual world users
1. Identify networks	Instructors should provide single entry points to virtual world projects (e.g., through a web page, handbook, or guide). Orientation and ongoing training should be tailored to the specific needs of STEM educators.	Identification and inclusion of multiple in-world (SL) and out-of-world networks. Create a group for specific course members.
2. Lurk	Visit, observe, and learn the culture and norms of the environment. Explore resources and tools that facilitate STEM learning (e.g., tools used in the KATE Project.).	Participate in in-world groups or networks, lurk on relevant listservs (e.g., Second Life Educators [SLED] list), attend talks, attend "building in SL" workshops.
3. Contribute	Begin interaction with others; help other learners as your comfort and skill level increases.	Volunteer or assist those hosting in-world events with marketing, traffic coordination; serve as a docent; make contributions to networks.
4. Create	Design/deliver STEM instruction or design/build something in-world (start small, sandbox areas are ideal for building. Include prior modeling of STEM lessons, before preservice teachers lead lessons.).	Participate in group builds (e.g., exhibit); design your own network around a given topic.
5. Lead	Establish groups in classes or special-interest groups of like-minded peers on relevant topics; host STEM guest speakers.	Create in-world groups on a special-interest topic(s); leverage partnerships among multiple group members to promote new learning opportunities and knowledge development.

ADDITIONAL SUGGESTIONS AND RESOURCES

A good starting place for exploring a variety of uses of *Second Life* is to connect with professional groups that are composed of experienced and novice SL users. Several organizations host in-world conferences and ongoing SL workshops and events. For example, the following groups offer outstanding practitioner and researcher events in SL and other virtual worlds:

- Applied Research in Virtual Environments (ARVEL SIG): An American Education Research Association (AERA) SIG for educators, scholars, and practitioners researching computer-mediated 3-D communities.
- International Society for Technology in Education, Special Interest Group Virtual Environments (SIGVE)
- EdTech Island and CAVE (Community of Academic Virtual Educators) Island
- Virtual Worlds Best Practices in Education Conference

Visiting *Second Life* locations outside of STEM areas can also expand your thinking in terms of what's possible and what can be done within this 3-D immersive environment (including seeing models of various learning affordances). For example, it may be valuable for preservice teachers or other stakeholders to attend informal learning or fine arts events as you get started (e.g., concerts, museums, art exhibitions). Partnerships and support are also important considerations for instructors or instructional design teams that are designing experiences for preservice teachers. Seek out experienced SL users in your local setting or within familiar organizations that are using virtual worlds in various teaching and learning contexts; they may be potential partners or offer just-in-time technical support as you are getting started. Finally, the Second Life in Education (SLED) listserv is an extremely active list. You might consider joining this list to connect with other educators on a daily basis. I recommend using an alternate email address, since myriad messages are posted daily.

REFERENCES

American Association for the Advancement of Science (AAAS). (2000). *Middle grades mathematics textbooks: A benchmarks-based evaluation.* Washington, DC: Author.

Annetta, L., & Park, J. (2006). Video games in science: A model for students and teachers creating 3-D role playing games. In C. Crawford et al. (Eds.), *Proceedings of the Society for Information Technology & Teacher Education international conference* (pp. 3950–3956). Chesapeake, VA: AACE.

Barab, S., Thomas, M., Dodge, T., Carteaux, R., & Tuzun, H. (2005). Making learning fun: Quest Atlantis, a game without guns. *Educational Technology Research and Development, 53*(1), 86–107.

Barr, D., Harrison, J., & Conery, L. (2011). Computational thinking: A digital age skill for everyone. *Learning & Leading with Technology, 38*(6), 20–52.

Bartle, R. (2004). *Designing virtual worlds.* Indianapolis, IN: New Riders.

Bayer Corporation. (2010). *Bayer facts of science education XIV: Female and minority chemists and chemical engineers speak about diversity and underrepresentation in STEM.* Pittsburgh, PA: Campos Inc.

Bell, M. (2001). Online role-play: Anonymity, engagement and risk. *Education Media International, 38*(4), 251–260.

Bell, M. W. (2008). Toward a definition of "virtual worlds." *Journal of Virtual Worlds Research, 1*(1), 1–5.

Brown, I. A., Davis, T. J., & Kulm, G. (2011). Preservice teachers' knowledge for teaching algebra for equity in the middle grades: A preliminary report. *Journal of Negro Education, 80*(3), 266–283.

Caprotti, O., & Seppala, M. (2007, October). Mathematics education in Second Life. In *Proceedings of the 6th Open Classroom Conference,* Stockholm, Sweden.

Carnegie Foundation. (2010). *Standard listings.* Retrieved from http://classifications.carnegiefoundation.org/lookup_listings/institution.php

Castronova, E. (2004). *Synthetic worlds.* Chicago, IL: Chicago University Press.

Cheng, H-J., Zhan, H., & Tsai, A. (2010). Integrating Second Life into a Chinese language teacher training program: A pilot study. *Journal of Technology and Chinese Language Teaching, 1*(1), 31–58.

Dalgarno, B., & Lee, M. J. (2010). What are the learning affordances of 3-D virtual environments? *British Journal of Educational Technology, 41*(1), 10–32. doi:10.1111/j.1467-8535.2009.01038.x

Davis, T. (2012a, April). Preservice teachers' and graduate research assistants' perceptions of classroom simulations and experiences in Second Life. In G. Kulm (Chair), *Preservice teachers' knowledge for teaching algebra: A preliminary report.* Research symposium conducted at the annual meeting of the National Council of Teachers of Mathematics, Philadelphia, PA.

Davis, T. J. (2012b). Mathematics learning in virtual worlds: Affordances and accessibility. In C. R. Reynolds, K. J. Vannest, & E. Fletcher-Janzen (Eds.), *Encyclopedia of special education: A reference for the education of children, adolescents, and adults with disabilities and other exceptional individuals* (4th ed., in press). Hoboken, NJ: John Wiley and Sons.

Dawley, L. (2009). Social network knowledge construction: Emerging virtual world pedagogy. *On the Horizon, 17*(2), 109–121.

Dickey, M. D. (2005). Three-dimensional virtual worlds and distance learning: Two case studies of active worlds as a medium for distance education. *British Journal of Educational Technology, 36*(3), 439–451.

Dillenbourg, P., Schneider, D. K., & Synteta, P. (2002). Virtual learning environments. In A. Dimitracopoulou (Ed.), *Proceedings of the 3rd Hellenic conference "Information & Communication Technologies in Education"* (pp. 3–18). Athens, Greece: Kastaniotis Editions.

Flores, A. (2007). Examining disparities in mathematics education: Achievement gap or opportunity gap? *The High School Journal, 91*(1), 29–42.

Freeman, M. A., & Capper, J. M. (1999). Exploiting the web for education: An anonymous asynchronous role simulation. *Australian Journal of Educational Technology, 15*(1), 95–116. Retrieved from http://www.ascilite.org.au/ajet/ajet15/freeman.html

Gao, F., Noh, J. M., & Koehler, M. J. (2008). Comparing student interactions in second life and face-to-face role-playing activities. *Technology and Teacher Education Annual, 19*(4), 2033.

Goodband, J., Bhakta, R., & Lawson, D. (n.d.). *Exploiting new technologies in mathematics support.* Centre for Excellence in Mathematics and Statistics Support, Faculty of Engineering and Computing, Coventry University, UK.

Gregory, S., & Masters, Y. (2010, March 2–6). *Six hats in second life: Enhancing preservice teacher learning in a virtual world.* Paper presented at the International Conference for Teaching and Learning with Technology (iCTLT), Singapore.

Harrell, S. V., Abrahamson, D., Morgado, L., Valcke, M., Vansteenburugge, H., & Rosenbaum, E. (2008, June 24–28). Virtually there: Emerging design for STEM teaching and learning in immersive online 3D microworlds. *In proceedings of the 8th international conference for the Learning Sciences* (pp. 383–391), Utrecht, The Netherlands.

Jauregi, K., Canto, S., de Graaff, R., Koenraad, T., & Moonen, M. (2011). Verbal interaction in Second Life: Towards a pedagogic framework for task design. *Computer Assisted Language Learning, 24*(1), 77–101.

Jenkins, H., Clinton, K., Purushotma, R., Robinson, A. J., & Weigel, M. (2006). *Confronting the challenges of participatory culture: Media education for the 21st century.* Chicago, IL: MacArthur Foundation, with permission from Henry Jenkins.

Jones, K. (1982). *Simulations in language teaching.* Cambridge, UK: Cambridge University Press.

Kaufmann, H., & Schmalstieg, D. (2003). Mathematics and geometry education with collaborative augmented reality. *Computer & Graphics, 27,* 339–345.

Kluge, S., & Riley, L. (2008). Teaching in virtual worlds: Opportunities and challenges. *Issues in Informing Science & Information Technology, 5,* 127–135.

Knowledge for Algebra Teaching for Equity. (2012). *Home.* Retrieved from http://kate.tamu.edu

Kulm, G., & Capraro, R. M. (2008). Textbook use and student learning of number and algebra ideas in middle grades. In G. Kulm (Ed.), *Teacher knowledge and practice in middle grades mathematics* (pp. 255–272). Rotterdam, The Netherlands: Sense.

Mahon, J., Bryant, B., Brown, B., & Kim, M. (2010). Using Second Life to enhance classroom management practice in teacher education. *Educational Media International, 47*(2), 121–134.

Malamos, A., Mamakis, G., Sympa, P., Kotanitsi, E., Crespo, A. J. G., & Lopez, A. Z. (2008, July 22–24). Technical aspects in using X3D in virtual reality mathematics education (EViE-M platform). In *Proceedings of 5th WSEAS/IASME International Conference on Engineering Education* (pp. 284–291). Heraklion, Crete Island, Greece.

McComas, J., Pivik, J., & Laflamme, M. (1998). Current users of virtual reality for children with disabilities. In G. Riva (Ed.), *Virtual environments in clinical psychology and neuroscience: Methods and techniques in advanced patient therapist interaction* (pp. 161–169). Amsterdam, The Netherlands: IOS.

Merchant, Z., Goetz, E., Keeney-Kennicutt, W., Kwok, O., Cifuentes, L., & Davis, T. (2012). The features of desktop 3D virtual reality environments, learner characteristics, and college chemistry instruction. *Computers & Education,* (in press).

Minocha, S., & Reeves, A. J. (2010). Design of learning spaces in 3D virtual worlds: An empirical investigation of Second Life. *Learning, Media and Technology, 35*(2), 111–137.

Moreno, R., & Ortegano-Layne, L. (2008). Do classroom exemplars promote the application of principles in teacher education? A comparison of videos, animations, and narratives. *Educational Technology Research and Development, 56*(4), 449–465.

Murfin, B. (2001). A case study of math and science teacher education in a collaborative virtual learning environment. *Journal of Computers in Mathematics and Science Teaching, 20*(4), 405–425.

National Council of Teachers of Mathematics. (2000). *Principles and standards of school mathematics.* Reston, VA: Author.

Ondrejka, C. (2008). Education unleashed: Participatory culture, education, and innovation in *Second Life.* In K. Salen (Ed.), *The ecology of games: Connecting youth, games, and learning* (pp. 229–252). The John D. and Catherine T. MacArthur Foundation Series on Digital Media and Learning. Cambridge, MA: MIT Press.

Scarcella, R., & Oxford, R. L. (1992). *The tapestry of language learning: The individual in the communicative classroom.* Boston, MA: Heinle & Heinle.

Thompson, S. (2008). Teaching in a virtual setting. In L. Bell & R. B. Trueman (Eds.), *Virtual worlds, real libraries: Librarians and education in Second Life and other multi-user virtual environments* (pp. 165–172). Medford, NJ: Information Today.

Tompkins, P. K. (1998). Role playing/simulation. *The Internet TESL Journal, 4*(8), 143–150.

Wang, Q., & Woo, H. L. (2007). Comparing asynchronous online discussions and face-to-face discussion in a classroom setting. *British Journal of Educational Technology, 38*(2), 272–286.

Wright, W. (1989). *SimCity* [Computer software]. Moraga, CA: Maxis.

CHAPTER 9

USING JUST-IN-TIME PD TO TECHNOLOGICALLY PREPARE HIGH SCHOOL STEM TEACHERS

J. Eli Crow
Teresa J. Kennedy
Michael R. L. Odell
University of Texas at Tyler & Ingenuity Center

John D. Ophus
University of Northern Iowa

Jason T. Abbitt
University of Miami (Ohio)

BACKGROUND

Using a technological, pedagogical, and content knowledge (TPCK) professional development (PD) model has shown promise in narrowing the achievement gaps between majority and minority student groups, including students labeled economically disadvantaged.

Improving Urban Schools, pages 143–157

The model incorporates a number of individual best practices found in the PD research literature as well as promising practices from practitioners, including

1. Is job-embedded;
2. Is "just in time," immediately relevant to teacher's content knowledge;
3. Uses technology as an "on-ramp" for pedagogy and content-based discussions;
4. Is supported by high quality coaching; and
5. Is supported by positive administrative reinforcement.

The model focuses on the introduction, integration, and implementation of learning technologies in the classroom. The advantage of utilizing technology as the focus of the training is that it allows for meaningful discussions of content and pedagogy in a nonthreatening manner.

Professional development occurs throughout the year and during the school day. To increase teacher buy-in, the PD focus is the upcoming curriculum to be taught during the next few weeks. Teachers work in teams to learn the technology and practice of teaching each other using the technology. This allows teachers to experience how students may have difficulties. Teachers discuss content and their practice. Instructional coaches are available to assist teachers with implementation both during the PD as well as in the classroom afterwards. School administrators also participate in the PD so they know what to look for during walk-through observations and so they can reinforce approximate success by the teacher. Implementation of the model has led to significant improvements in the achievement of all student groups. Of particular emphasis is the closing of the gaps between minority and low-socioeconomic student groups as compared to traditionally higher achieving students.

Literature Review

There are two significant goals that the model was designed to address. The first was to improve teacher instruction. The second was to improve student achievement on state assessments, particularly attending to closing the achievement gap between underrepresented groups and majority students.

Closing the achievement gap is described by the National Assessment of Educational Progress (NAEP) as when "one group of students outperforms another group and the difference in average scores for the two groups is statistically significant" (Institute of Education Sciences, 2011). The concept of the achievement gap dates back to the passage of the Civil Rights Act of 1964. In 1966, the U.S. Department of Health, Education, and Welfare commissioned James Samuel Coleman to investigate the equality of education available to children of different races, color, national origin, and religion (Coleman et al., 1966). The result was the report "Equality of Educational Opportunity," in which Coleman and colleagues reported that even though funding in schools for Black students was nearly on par with that of schools for White students, achievement between the two groups was substantially different.

It can be argued that Texas has had as big a role in attempting to close the achievement gap as any state, and with good reason. Data from the 2010 U.S. Census reveals that 11.8% of the population identified themselves as African American and 37.6% of Hispanic or Latino origin (U.S. Census Bureau, 2010). Yet despite making up a significant portion of the population, both groups lag significantly behind White students, according to the Texas Assessment of Knowledge and Skills (TAKS). Data from 2011 revealed that only 63% of tenth-grade African American students were at or above the state's standard for science proficiency, whereas 71% of Hispanic students and 84% of White students met the standard. Student results for mathematics were similar: African American, 64%; Hispanics, 69%; and White, 88% (Texas Education Agency, 2011).

Undeniable progress has been made nationally. Between 1971 and 1988, African American 17-year-olds increased an average of 0.6 standard deviations on the NAEP, while White 17-year-olds increased less than a tenth of a standard deviation (Jencks & Phillips, 1998). Between 1996 and 2000, the number of students scoring "below basic" on the Texas Academic Assessment of Skills (TAAS) decreased for African Americans from 69% to 60% and for Hispanic students from 58% to 41% (Linton & Kester, 2003).

Though it seemed that Texas was making significant progress in closing the gap, evidence suggests that the problem may have only been masked by weaknesses in the design of the TAAS tests. When the TAKS was introduced in 2004, the gap reappeared. The cumulative results of the 2004 Science TAKS tests indicated that only 38% of African American students, 41% of Hispanic students, and 39% of economically disadvantaged students met the state standard. When compared to White students, 73% of which met the standard, the gap was clearly apparent.

While many possibilities exist that may contribute to the problem, teacher quality is undoubtedly tied to closing the gap in science, technology, engineering, and mathematics (STEM) learning. Studies have linked student high-stakes scores and teacher quality (Boyd, Lankford, Loeb, Rockoff, & Wyckoff, 2007; Dobbie & Fryer, 2009; Hanushek & Rivkin, 2006). This chapter discusses the best practices in TPCK PD.

TECHNOLOGICAL, PEDAGOGICAL, AND CONTENT KNOWLEDGE (TPCK)

The relationship and intersections among knowledge of pedagogy, content, and technology has been explored by researchers in multiple, and often parallel, ways that extend or contextualize the concept of pedagogical content knowledge (PCK). In an article describing PCK, Shulman (1986) suggested that there is a long history of exploration and inquiry into teacher knowledge of pedagogy and content, and that much of this history focused on the connection between knowledge of pedagogy and content disciplines rather than viewing these forms of knowledge as separate and distinct. Shulman further articulated a definition of the PCK as a "particular form of content knowledge that embodies the aspects of content most germane to its teachability" (p. 9) and that this knowledge also includes an understanding of both how learning in a content area may be easy or difficult and the influence of preconceptions and misconceptions on learning in a particular content discipline.

The TPCK framework represents the forms of knowledge that support effective and innovative integration of technology into classroom teaching practices (see Figure 9.1) (Mishra & Koehler, 2006). Seven forms of knowledge composed of three primary knowledge domains (pedagogical, content, and technological) and four integrated domains representing the intersections among the three primary domains. These seven knowledge domains are summarized as follows:

1. Pedagogical Knowledge (PK): Knowledge of teaching and learning, including teaching methods, classroom management, instructional planning, assessment of student learning, and so on.
2. Content Knowledge (CK): Knowledge about the subject matter that is being taught, such as science, language arts, social studies, and so on.
3. Technology Knowledge (TK): A rapidly evolving knowledge base that includes knowledge of technology for many purposes including information processing, communications, and problem solving that is focused on the use of technology in both work and daily life.
4. Pedagogical Content Knowledge (PCK): A blended form of knowledge of the pedagogies that are applicable, appropriate, and effective when teaching a given subject matter.
5. Technological Content Knowledge (TCK): Knowledge of technology; a relevant content discipline that includes knowledge of how technology supports real-world applications within a content discipline.
6. Technological Pedagogical Knowledge (TPK): Knowledge of the various ways in which technology influences teaching and learning,

including the affordances and constraints of technology for pedagogical designs and strategies.
7. Technological Pedagogical Content Knowledge (TPCK): Knowledge of the complex interaction among the principle knowledge domains (content, pedagogy, technology).

As Mishra and Koehler indicate, "a conceptually based theoretical framework about the relationship between technology and teaching can transform the conceptualization and the practice of teacher education, teacher training, and teachers' professional development" (2006, p. 1019).

Although the TPCK framework has been both the focus of and the lens for recent research efforts, there have also been several parallel efforts that arrived at similar ideas that highlight various facets of technology integration. Effective technology integration requires much more than knowledge of technology for personal use; it requires an integrated knowledge of the transformative potential of the blended knowledge of pedagogy, content, and technology.

Using the definition of PCK as "knowledge of general pedagogical strategies and the ability to apply those strategies", the authors extended the idea, incorporating the use of technological" (Margerum-Leys & Marx, 2002, p. 430). There is an exchange of technology knowledge between student teacher/mentor pairs in which teachers learned about technology from student teachers and then integrated this technology into classroom

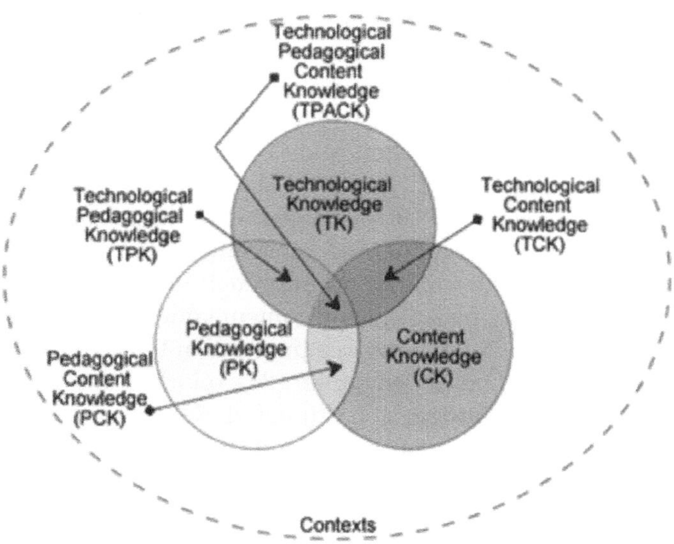

Figure 9.1 Technological pedagogical content knowledge (Mishra & Koehler, 2006).

practice. There is a complex interplay between technology and pedagogy that suggests that developing knowledge about specific technologies leads to a deeper exploration of the application of technology in the classroom.

TPCK and Professional Development

As the concept of PCK has morphed into various models, the TPCK framework has become the most prominent expansion of Shulman's PCK model. However, the contexts to which the TPCK framework has been applied as well as the purposes that it serves have been varied and numerous. Among the many variations in the application of the TPCK framework are the uses as either a lens for research about the development of teacher knowledge (Archambault & Oh-Young, 2009; Burgoyne, Graham, & Sudweeks, 2010; Cox & Graham, 2009; Graham, Burgoyne, & Borup, 2010; Koehler & Mishra, 2005; Schmidt et al., 2009a, 2009b) and a frame to model preparation of both inservice and preservice teachers (Brush & Saye, 2009; Figg & Jaipal, 2009; Hardy, 2010; Harris et al., 2010; Harris, Mishra, & Koehler, 2009; Hofer & Harris, 2010; Hofer et al., 2009; Koehler, Mishra, Hershey, & Peruski, 2004; Mishra & Koehler, 2007; Niess, 2006, 2007; Niess et al., 2009; Polly & Barbour, 2009; Reeve, 2008; Schmidt, Sahin, Thompson, & Seymour, 2008).

Within the context of teacher PD, there has been a more narrow focus on the practical application of teacher knowledge in the TPCK domains. There is often a focus on progressing toward TPCK in which teachers are developing an understanding of the interconnectedness of technology, pedagogy, and content knowledge. However, there appears to be different starting points in this progression. Neiss (2007) described a program in which the goal was to "engage the teachers in developing their TPCK for teaching mathematics with spreadsheets" (p. 2241). The authors of that study suggested that the development of TPCK was supported by "opportunities to experience learning about the technologies' capabilities" (p. 2244). The process by which teachers progress toward an integrated understanding of the intersection of technology, pedagogy, and content knowledge through various stages is more specifically discussed as a progression through five stages: (a) recognizing, (b) accepting, (c) adapting, (d) exploring, and (e) advancing (Niess et al., 2009). The progression through these stages requires ongoing rethinking of the relationship of content and pedagogies as well. In this way, inservice teachers, who may have developed a rich understanding of pedagogy through their experience, become engaged in the discussion of pedagogy and content through the catalyst of exploring technology.

A focus on "content-based learning activities... rather than the affordances and constraints of educational technologies" may yield more positive results (Hofer et al., 2009, p. 575). An example would be the Learning Activity Taxonomy, in which collections of common types or learning activities are organized by content area. This approach is further described as "grounded" in that the technologies are then connected to the subject matter through content-specific pedagogies. As such, this approach contrasts with the one described by Neiss (2009) in which planning for technology integration begins with an exploration of the intersection of pedagogy and content and then integrates technologies that are appropriate for this unique instructional context. Although the starting point for each approach may be distinct, both share a goal of fostering critical thinking within the blended TPCK domain. As such, it is important to consider the potential ways in which a PD program may employ a strategy that uses one of the areas of teacher knowledge as an "on-ramp" to the other two primary domains.

Application of UT Tyler Ingenuity Center TPCK Model

Following the 2008 school year, a high school was selected for the application of the TPCK model. The school participated in the Texas Regional Collaboratives for Science, a project that allowed for financial backing as well as the level of commitment necessary to fully implement the model. Teachers were provided with a substitute one day each month during the school year. Teachers were gathered by department and engaged in a full day of on-site PD. Each department (Biology, Chemistry, and Physics) met separately and focused specifically on their content. Every teacher was required to attend, and attendance was strictly enforced by the administration. The PD was always structured in the same manner. Technological tools were used to introduce concepts that would be covered by the teachers in the upcoming month. Teachers were allowed to explore both the technology and the concept and were engaged in a TPCK discussion. The discussion asked the question, "How do we use the technology to teach this concept?" In this manner, the focus was on the technology, but the conversation quickly moved to the content and the surrounding pedagogy, thus completing the TPCK discussion model.

Participants

Table 9.1 shows the demographics for the enrollment of the school of choice. Each student in the school was required to take a state exam in science at the end of the tenth-grade year and again at the end of the eleventh-grade year. The test was designed as an exit-level administration for the eleventh-grade year, and students were required to meet a minimum

TABLE 9.1 Target School Enrollment Demographics: 2008–2010

	2008	2009	2010
Enrollment	2,661	2,591	2,637
African American	26.0%	26.0%	27.1%
Hispanic	18.1%	18.8%	20.2%
White	53.0%	52.3%	49.9%
Economically disadvantaged	35.5%	34.4%	38.6%

standard in order to graduate. The schools were held accountable for the sum of all tested grades.

Analysis

A review of the results for the entire state reveals the following pattern as displayed in Figure 9.2.

The minimum acceptable standard for a school is shown in bold at the bottom, and the sum of all schools has kept pace with the increasing standard. A considerable gap is evident between minority and economically disadvantaged students and that of White students.

A similar review of the scores from the target school can be seen in Figure 9.3.

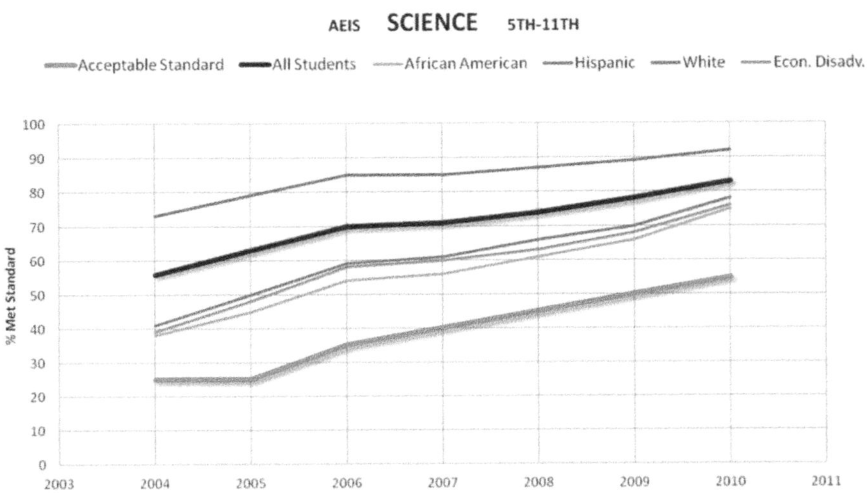

Figure 9.2 Science achievement results: Texas academic excellence indicator system.

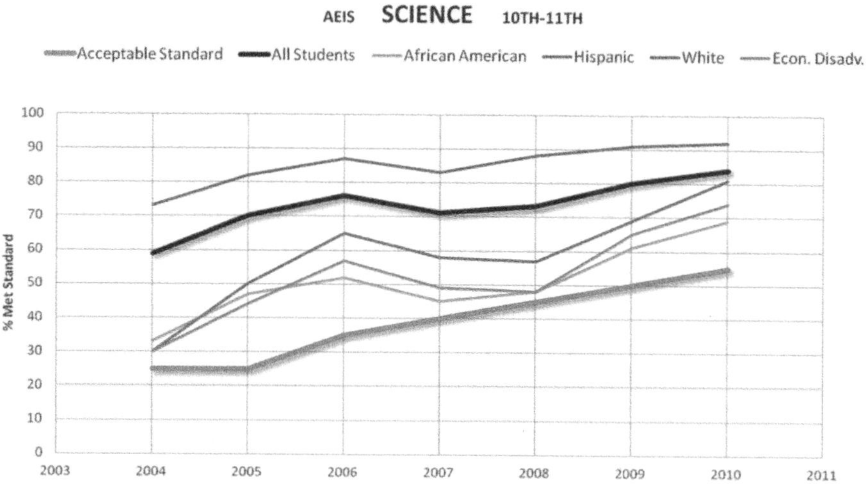

Figure 9.3 Target school science achievement results.

The graph depicts a school with declining scores and a broadening of the gap from 2006 to 2008. It is important to note that the state altered the inclusion of special-education students during these years. Prior to these years, special-education students were not included in the accountability scores. In the 2007 school year, the state began requiring schools to include these students in accountability calculations. Also noted is the distinct increase in the scores of minority and economically disadvantaged students from 2008 to 2010. Though all students showed growth, the greatest gains were made in these groups. Also shown is a clear closing of the gap between student groups.

The change in the gap between the higher performing White students and the lower performing minority students can be seen in Table 9.2. The figure indicates that the gaps were narrowed considerably. In 2008, prior to the implantation of the TPACK model, the percentage of economically disadvantaged students who met the state standard was 40 points lower

TABLE 9.2 Achievement Gaps Between Student Groups

	2008		2010	
Student group	**% met**	**gap**	**% met**	**gap**
White	88	0	92	0
African American	48	40	69	23
Hispanic	57	31	81	11
Economically disadvantaged	48	40	74	18

than the percentage of White students who met the standard. Following the implementation of the model, this gap was narrowed to only 18 points. Hispanic students showed the largest narrowing as the gap moved from 31 points to only 11, and African American students also improved considerably. In 2008, only 48% of African American students were meeting the state standard, but in 2010, some 69% of students in this group were successful. This represents a 21% increase, and the gap between these students and their higher performing peers narrowed from 40 points to 23.

An independent means *t*-test was performed to test for significance in the change in scores from the 2008 to 2010 school year. All student scores from the 2008 science TAKS tests from the sample school were compared with all of the scores from the 2010 science TAKS tests from the same school. The students from the sample in 2010 had higher scores (raw scores $M = 40$, $SD = 9.45$; scale score $M = 2{,}241$, $SD = 170.85$) than the students in the sample from 2008 (raw scores $M = 37.5$, $SD = 10.8$; scale score $M = 2{,}186$, $SD = 176.13$). The difference was statistically significant for raw scores $t(2{,}324) = -5.71$, $p < .0001$ d = .02, as well as scale scores $t(2{,}324) = -7.21$, $p < .0001$, d = .001.

CONCLUSIONS

The application of the TPCK model has been replicated several times since these data were collected. Those applications have been in diverse settings including middle school science, high school mathematics, and middle school cross-curricular teams. What has emerged in those replications is a clear pattern that the TPCK model is intended to be implemented as a holistic model and that significant interaction occurs between the components.

The TPCK model shows significant promise as a PD paradigm. By combining job-embedded, just-in-time training that utilizes technology as an "on-ramp" for high-level PCK conversations with coaching and positive reinforcement, significant improvements in science instruction can be obtained. These improvements have positive impacts on all student groups but are particularly effective in closing the gaps between majority and minority students.

RECOMMENDATIONS

School leaders facing increasing standards and limited resources are often faced with the choice: Do I use instructional time to develop my teachers, or do I maximize the time teachers spend in front of students? Though this seems like a difficult trade-off, we believe that the question is the wrong one to ask. A better question would be, Am I satisfied with the current level of instruction and student performance? If the answer is no, then something must change. Simply having teachers do more of what they are already doing is not going to significantly improve student performance. If we want student performance to improve, teaching must improve, and this will take time and resources. The good news is this investment in teachers and teaching will more than compensate for any instructional time lost in the process.

Recommendations for School Leaders

The TPCK framework is not a menu from which to choose the parts that you like and pass on the ones you do not, or think you cannot afford. The TPCK framework is a model for staff development with critical components that work to leverage the gains of the other components, such that the whole becomes more than the sum of the parts. If an administration wants to improve student performance, they must make time to improve the skills of their faculty. Whether a school is the highest performing in the state, or the lowest, teachers need opportunities to improve, and students deserve the most effective teachers. Implementing the TPCK framework is not inexpensive. It requires a commitment to the tools, the time, the manpower, and the personal commitment of school leaders, but the results speak for themselves.

If a school is considering implementing a PD initiative based on the TPCK model, consider the following. First, the teachers must have ready access to the equipment. We recommend that teachers receive a classroom set of technology tools. Many schools have attempted to reduce cost by purchasing a set for the department to share. Unfortunately, this is rarely successful. The goal is for the technology to become part of the everyday learning environment, which can happen only if the teacher and the students can access the tools daily. For the teacher, there is nothing worse than going to the trouble of planning a new lesson based on new tools only to find that the class that had it the day before left it in such disarray that there is no chance of success. The teachers simply need ownership of the tools.

Second, we all have personal lives outside of work, and our teachers do too. Given the amount of time and effort that they pour into their work, it is simply unreasonable to think that there will be anything left at the end of the day. Teachers cannot be asked to do more. If we expect real change, then we need to provide PD at the time when they are most receptive to it, and it cannot be a one-time event. Professional development that creates real change is ongoing and job embedded. Additionally, it must be relevant. It is foolish to think that one could provide PD in August for something that will not be implemented until the following March. Professional development that creates real change is just-in-time.

Finally, we must recognize that teachers need support. Oftentimes, we are asking teachers to do something different from the way they have always done it. It should be obvious that some amount of apprehension will exist. Professional development providers need to make sure that teachers have a support structure in place—a coach that will cheer them on and help them when they fall short. The administrator has a role in this and must be a part of it, but experience indicates that the administrator who devotes 12–15 full days to a project like this is rare if not nonexistent. Thus, teachers need a coach; someone who they can trust to be there.

Many districts attempt to do this by hiring one person for multiple campuses. Unfortunately, two things occur, which leads us to discourage districts from this type of approach. The first is on the teacher side. Inevitably, something goes wrong when the coach is not there. The whole point of a coach is dependability; they cannot be in two places at once. The second is on the side of the coach. Anyone who has been in this position can tell you that you have your favorites. As much as a coach tries to be equitable, one simply cannot be completely loyal to everyone. When coaches are on a campus full-time, they own the results and are more dedicated to change. Teachers need someone who can be there for them, when they are needed, every time they are needed.

Recommendations for Researchers

The TPCK provides a framework for PD that holds promise for improvements in teaching and learning, but more work needs to be done. Our research focused on the science classroom, a content area that lends itself well to a technology-driven, investigative approach. Science teachers, by the nature of their training, are inquisitive and often open to new ideas.

In addition to the portability of the model, more research needs to be done in diverse settings and considering longitudinal studies. What differences might we see in high performing schools? Could the TPCK model be used against the more rigorous measures of advanced placement and international baccalaureate testing? One might also consider what happens over time. Do the results tend to fade as teachers become more familiar with the technologies, or does the professional developer continue to push the envelope with new technologies? The answers to these and other questions will help school leaders continue to fine tune the PD of teachers and ensure that students are able to continue to meet the ever-increasing challenges of tomorrow.

REFERENCES

Archambault, L., & Oh-Young, C. (2009). Putting the T in PCK: Exploring the nature of the TPCK framework among K–12 online educators using a web-based survey. In I. Gibson, R. Weber, K. McFerrin, R. Carlsen, & D. A. Willis (Eds.), *Proceedings of the Society for Information Technology & Teacher Education international conference 2009* (pp. 4008–4014). Chesapeake, VA: AACE.

Boyd, L., Lankford, H., Loeb, S., Rockoff, J., & Wyckoff, J. (2007). The narrowing gap in New York City teacher qualifications and its implications for student achievement in high-poverty schools. *Teacher Policy Research.* Retrieved from http://www.teacherpolicyresearch.org/ResearchPapers/tabid/103/Default .aspx

Brush, T., & Saye, J. (2009). Strategies for preparing preservice social studies teachers to integrate effectively: Models and practices. *Contemporary Issues in Technology and Teacher Education, 9*(1), 46–59.

Burgoyne, N., Graham, C. R., & Sudweeks, R. (2010). The validation of an instrument measuring TPCK. In C. Crawford, D. A. Willis, R. Carlsen, I. Gibson, K. McFerrin, J. Price, & R. Weber (Eds.), *Proceedings of the Society for Information Technology & Teacher Education international conference 2010* (pp. 3787–3794). Chesapeake, VA: AACE.

Coleman, J. S., Campbell, E. Q., Hobson, C. J., McPartland, J., Mood, A. M., Weinfeld, F. D., & York, R. L. (1966). *Equality of educational opportunity.* Washington, DC: U.S. Department of Education, Government Printing Office.

Cox, S., & Graham, C. (2009). Diagramming TPCK in practice: Using an elaborated model of the TPCK framework to analyze and depict teacher knowledge. *TechTrends, 53*(5), 60–69.

Dobbie, W., & Fryer, R. (2009). *Are high quality schools enough to close the achievement gap? Evidence from a social experiment in Harlem.* NBER Working Paper 15473.

Figg, C., & Jaipal, K. (2009). Unpacking TPCK: TPK characteristics supporting successful implementation. In I. Gibson, R. Weber, K. McFerrin, R. Carlsen, & D. A. Willis (Eds.), *Proceedings of the Society for Information Technology & Teacher Education international conference 2009* (pp. 4069–4073). Chesapeake, VA: AACE.

Graham, C. R., Burgoyne, N., & Borup, J. (2010). The decision-making processes of preservice teachers as they integrate technology. In C. Crawford, D. A. Willis, R. Carlsen, I. Gibson, K. McFerrin, J. Price, & R. Weber (Eds.), *Proceedings of the Society for Information Technology & Teacher Education international conference 2010* (pp. 3826–3832). Chesapeake, VA: AACE.

Hanushek, E. A., & Rivkin, S. G. (2006). *School quality and the Black-White achievement gap* (Working Paper No. 12651). Cambridge, MA: National Bureau of Economic Research.

Hardy, M. (2010). Enhancing preservice mathematics teachers' TPCK. *Journal of Computers in Mathematics and Science Teaching, 29*(1), 73–86.

Harris, J., Hofer, M., Blanchard, M., Grandgenett, N., Schmidt, D., van Olphen, M., & Young, C. (2010). "Grounded" technology integration: Instructional planning using curriculum-based activity type taxonomies. *Journal of Technology and Teacher Education, 18*(4), 573–605.

Harris, J., Mishra, P., & Koehler, M. (2009). Teachers' technological pedagogical content knowledge and learning activity types: Curriculum-based technology integration reframed. *Journal of Research on Technology in Education, 41*(4), 393–416.

Hofer, M., & Harris, J. (2010). Differentiating TPCK development: Using learning activity types with inservice and preservice teachers. In C. Crawford, D. A. Willis, R. Carlsen, I. Gibson, K. McFerrin, J. Price, & R. Weber (Eds.), *Proceedings of the Society for Information Technology & Teacher Education international conference 2010* (pp. 3857–3864). Chesapeake, VA: AACE.

Hofer, M., Harris, J., Blanchard, M., Grandgenett, N., Schmidt, D., van Olphen, M., & Young, C. (2009). Operationalizing TPCK for educators: The activity types approach to technology integration, part 1. In I. Gibson, R. Weber, K. McFerrin, R. Carlsen, & D. A. Willis (Eds.), *Proceedings of the Society for Information Technology & Teacher Education international conference 2009* (pp. 4099-4102). Chesapeake, VA: AACE.

Institute of Education Sciences. (2011). *National Assessment of Educational Progress.* Retrieved from http://nces.ed.gov/nationsreportcard/studies/gaps/

Jencks, C., & Phillips, M. (1998). *The Black-White test score bap.* Washington, DC: Brookings Institution.

Koehler, M., & Mishra, P. (2005). What happens when teachers design educational technology? The development of technological pedagogical content knowledge. *Journal of Educational Computing Research, 32*(2), 131–152.

Koehler, M., Mishra, P., Hershey, K., & Peruski, L. (2004). With a little help from your students: A new model for faculty development and online course design. *Journal of Technology and Teacher Education, 12*(1), 25–55.

Linton, T. H., & Kester, D. (2003, March 14). Exploring the achievement gap between White and minority students in Texas: A comparison of the 1996 and 2000 NAEP and TAAS eighth grade mathematics test results. *Education Policy Analysis Archives, 11*(10). Retrieved from http://epaa.asu.edu/epaa/v11n10/

Margerum-Leys, J., & Marx, R. (2002). Teacher knowledge of educational technology: A case study of student/mentor teacher pairs. *Journal of Educational Computing Research, 26*(4), 427–462.

Mishra, P., & Koehler, M. (2006). Technological pedagogical content knowledge: A framework for teacher knowledge. *Teachers College Record, 108*(6), 1017–1054.

Mishra, P., & Koehler, M. (2007). Technological pedagogical content knowledge (TPCK): Confronting the wicked problems of teaching with technology. In R. Carlsen, K. McFerrin, J. Price, R. Weber, & D. A. Willis (Eds.), *Proceedings of the Society for Information Technology & Teacher Education international conference 2007* (pp. 2214–2226). Chesapeake, VA: AACE.

Niess, M. (2006). Preparing preservice teachers to teach mathematics with technology: Developing a TPCK. In C. Crawford, R. Carlsen, K. McFerrin, J. Price, R. Weber & D. A. Willis (Eds.), *Proceedings of the Society for Information Technology & Teacher Education international conference 2006,* (pp. 3788–3795). Chesapeake, VA: AACE.

Niess, M. (2007). Developing teacher's TPCK for teaching mathematics with spreadsheets. In R. Carlsen, K. McFerrin, J. Price, R. Weber, & D. A. Willis (Eds.), *Proceedings of the Society for Information Technology & Teacher Education international conference 2007* (pp. 2238–2245). Chesapeake, VA: AACE.

Niess, M., Ronau, R. N., Shafer, K. G., Driskell, S. O., Harper, S. R., Johnston, C., . . . Kersaint, G. (2009). Mathematics teacher TPCK standards and development model. *Contemporary Issues in Technology and Teacher Education, 9*(1), 4–24.

Polly, D., & Barbour, M. (2009). Developing teachers' technological, pedagogical, and content knowledge in mathematics. In I. Gibson, R. Weber, K. McFerrin, R. Carlsen, & D. A. Willis (Eds.), *Proceedings of the Society for Information Technology & Teacher Education international conference 2009* (pp. 4128–4131). Chesapeake, VA: AACE.

Reeve, R. (2008). Technological pedagogical content knowledge and the context: An integrated model. In K. McFerrin, R. Weber, R. Carlsen, & D. A. Willis (Eds.), *Proceedings of the Society for Information Technology & Teacher Education international conference 2008* (pp. 5310–5312). Chesapeake, VA: AACE.

Schmidt, D., Baran, E., Thompson, A., Mishra, P., Koehler, M., & Shin, T. (2009a). Examining preservice teachers' development of technological pedagogical content knowledge in an introductory instructional technology course. In G. Ian, W. Roberta, M. Karen, C. Roger, & W. Dee Anna (Eds.), *Proceedings of the Society for Information Technology & Teacher Education international conference 2009* (pp. 4145–4151). Chesapeake, VA: AACE.

Schmidt, D., Baran, E., Thompson, A., Mishra, P., Koehler, M., & Shin, T. (2009b). Technological pedagogical content knowledge (TPCK): The development and validation of an assessment instrument for preservice teachers. *Journal of Research on Technology in Education, 42*(2), 123–149.

Schmidt, D., Sahin, E. B., Thompson, A., & Seymour, J. (2008). Developing effective technological pedagogical and content knowledge (TPCK) in preK-6 teachers. In K. McFerrin, R. Weber, R. Carlsen, & D. A. Willis (Eds.), *Proceedings of the Society for Information Technology & Teacher Education international conference 2008* (pp. 5313–5317). Chesapeake, VA: AACE.

Shulman, L. (1986). Those who understand: Knowledge growth in teaching. *Educational Researcher, 15*(2), 4–14.

Texas Education Agency. (2011). *Performance reporting: 2011 state AEIS report.* Retrieved from http://ritter.tea.state.tx.us/perfreport/aeis/2011/state.html

U.S. Census Bureau. (2010). *State and county quickfacts: Texas.* Retrieved from http://quickfacts.census.gov/qfd/states/48000.html

CHAPTER 10

GENDER EQUITY IN STEM

Increasing the Persistence of Females in the STEM K–12 Pipeline

Meredith Jones
Texas A & M University

INTRODUCTION

In recent years, increasing attention has been paid to the importance of gender equity in science, technology, engineering, and mathematics (STEM). Despite the tremendous progress that women have made in educational and occupational attainment, females continue to be underrepresented in the STEM fields. The gap between males and females is particularly striking in certain STEM disciplines such as physics, engineering, and computer science and in STEM leadership positions (Hill, Corbett, & St. Rose, 2010; Schiebinger, 2010).

There are several reasons for promoting STEM careers among females. First, there is the issue of social equity. All individuals, regardless of gender, should have the opportunity to pursue STEM education and occupations. Second, though STEM professionals make up a small minority of the workforce, they are critical to innovation and promoting economic competitive-

Improving Urban Schools, pages 159–167

ness (NAS, NAE, & IOM, 2005). There is also a need to build diversity in the STEM field. The lack of female researchers and innovators in STEM has been described as leading to "the exclusion of specific types of knowledge women develop and maintain as a consequence of gender roles" (Schiebinger, 2010, p. 3). Another reason for promoting STEM education and careers among women is to reduce the gender pay gap. The STEM field is characterized by higher wages than other fields, a smaller wage gap between male and female workers, and better job security. According to the U.S. Department of Commerce, women in STEM occupations earned 33% more than women in non-STEM occupations (Beede, Julian, Langdon, McKittrick, Khan, & Doms, 2011).

This chapter will focus on gender equity in STEM, beginning with an overview of females in the STEM pipeline, followed by a discussion of why a gender gap persists in STEM, and concluding with recommendations for educators.

Females in the STEM Pipeline

The school-to-career STEM pipeline begins very early in a child's education. Research has found that in elementary school, before any gender differences in participation in mathematics and science classes have emerged, children already associate mathematics more strongly with males than females (Cvencek, Greenwald, & Meltzoff, 2011). While there are significant individual differences, boys also have a higher perceived competence in mathematics compared with girls.

Middle school is another important stage along the STEM pipeline. Students' performance in mathematics courses is particularly critical, as it typically determines students' placement in high school mathematics (Kinzie, 2007). By early adolescence, male and female students are already expressing different attitudes toward STEM subjects. Girls report less positive attitudes toward science and a greater interest in pursuing people-oriented jobs. In contrast, boys report more positive attitudes toward science and express more interest in jobs that involve physical objects and abstract concepts (Stake, 2006).

High school is a critical stage in the STEM pipeline, a point at which students either lose interest in STEM and opt out of advanced classes in mathematics and science or remain in the STEM pipeline, pursing advanced courses in mathematics and science. After high school, students may opt out of the STEM pipeline, but rarely do students "opt in" to STEM (Muller, Stage, & Kinzie, 2001). One of the best predictors of whether or not high school students will go on to pursue a STEM degree is whether they take the most advanced courses in mathematics and science during high school.

While females take many advanced courses in STEM areas, they do not complete the *most* advanced courses at the same rate as male students (Tyson, Lee, Borman, & Hanson, 2007). Research shows that gender differences in mathematics achievement does not explain the participation differences in advanced mathematics courses (Watt, Eccles, & Durik, 2006).

College is another important stage along the STEM pipeline; however, by the time most students reach this point, previous school experiences have largely established whether or not they will pursue a STEM major. In the first year of college, approximately twice as many males as females report plans to major in a STEM field (i.e., 29% of males compared to just 15% of females). If biology is not considered in this data, then the gap becomes even wider, with 20% of male freshmen reporting plans to major in STEM fields compared to just 5% females (Hill et al., 2010). Research into whether males and females drop out of STEM majors at the same rate has found somewhat mixed results. Griffith (2010) found that women and minority students were slightly less likely to persist in STEM majors; however, other studies have found no differences in the retention of males and females in STEM majors (Blickenstaff, 2005).

The number of women who receive graduate degrees in STEM areas has significantly increased in the past few decades. In 2006, women earned approximately 50% of the doctoral degrees in biological and agricultural sciences, 30% of the doctoral degrees in mathematics, and 20% of the doctoral degrees in computer science, engineering, and physics. However, while this is important progress, the percentage of women receiving doctoral degrees in computer science has actually declined since the 1980s, showing that we cannot take progress for granted (Hill et al., 2010).

Women who make it through the STEM pipeline and enter the STEM workforce are in the minority. Despite making up almost half (48%) of the workforce, only 24% of workers in STEM occupations are female. In STEM occupations, the largest percentage of women are in jobs related to physical and life sciences (40%) with the least in engineering (14%) (Hill et al., 2010).

Why is There a Gender Gap in STEM?

There have been a number of explanations as to why women do not pursue STEM education and careers at the same rate as men. A common belief is that a gender gap in STEM persists because males have innately superior mathematics abilities and thus are more likely to pursue careers that require higher-level mathematics skills. However, research conducted over the past few decades has shown that while there are some gender differences in cognitive abilities, they are subtle and would not account for the widely disproportionate representation of males and females in STEM careers (Spelke, 2005). Another explanation has been that women are not interested in the long preparation and workplace pressures associated with a career in STEM. However, women are now completing medical school at roughly equivalent rates as men and law school at even greater rates than men, both of which require lengthy training and have significant workplace pressures (Sevo, 2009).

Today, most researchers believe that the persistent gender gap in STEM education and occupations is the result of a number of factors, which occur at several different levels (i.e., individual, family, school, workplace, and society; Burke & Mattis, 2007). The following section will provide a selective review of the factors that contribute to the persistent gender gap in STEM.

Early Experiences

Early experiences may contribute to gender disparities in STEM. Research has shown that there are significant gender differences in children's early mathematics environments. Chang, Sandhofer, and Brown (2011) examined gender biases in early number exposure among young children and found that male children received significantly more input related to numbers. For example, mothers were much more likely to use cardinal numbers (e.g., the number of items in a set) with sons. While this behavior may be unconscious among parents, it can be very powerful for young children in terms of influencing their interests and self-competence. As stated by Chang et al. (2011), "less conversation about numbers with girls may reduce their awareness of quantities and mathematical concepts in everyday life, which may decrease their later interests in related domains" (p. 446).

Gender Stereotypes

Stereotypes about females' abilities and interest in STEM and suitability for STEM careers provide another factor that contributes to gender differences in STEM. Many parents and teachers believe that males are more naturally gifted in mathematics and science. This was illustrated in a study in which parents who were asked to estimate their children's IQ scores in mathematics reported their sons' mathematics intelligence scores to be significantly higher than their daughters (Furnham, Reeves, & Budhani, 2002). Parents and teachers who endorse these beliefs can influence their children's self-confidence in mathematics and science and reduce their desire to pursue education and careers in STEM fields (Kurtz, Rowley, Harris-Britt, & Woods, 2008). Gender stereotypes about STEM may also discourage girls from pursuing STEM education or occupations as they perceive it as not being consistent with their gender role (Steinke et al., 2007).

Gender stereotypes may also hinder females' performance in mathematics and science through the experience of "stereotype threat." Stereotype threat refers to the experience of anxiety or fear in a situation in which there is a risk of confirming a negative stereotype about a group to which you belong. Females have been found to experience stereotype threat in situations in which they may "risk" confirming negative stereotypes about their mathematics or science abilities. Examples of situations in which females might experience stereotype threat typically include exams and class discussions dealing with STEM topics. Indeed, research has shown that when an examiner makes female students aware of their gender or somehow highlights the stereotype about males' "natural advantage" in mathematics, females perform more poorly on mathematics exams. Indeed, research bears out such worries concerning the effect of stereotype threat on female students during mathematics exams (Spencer, Steele, & Quinn, 1999). It is hypothesized that stereotype threat may explain the small but significant overrepresentation of males among the top scorers on the mathematics portion of the SAT.

Classroom Experiences

Gender differences in students' experiences in STEM classrooms also contribute to gender disparities in STEM (Bachman, Hebl, Martinez, & Rittmayer, 2009). Research has shown that male students typically receive more attention and encouragement from their mathematics teachers than female students (Eccles, Freedman-Doan, Frome, Jacobs, & Yoon, 2000). Similar findings have been found in science classrooms, with research indicating that girls receive less attention from science teachers and have fewer opportunities to engage with science equipment (Brotman & Moore, 2008). Science content, particularly in the physical sciences, has also been found to be less applicable to the real-life experiences and interest of girls compared to boys (Burke & Mattis, 2007).

Spatial Skills

Spatial skills constitute another factor that has been identified as contributing to the gender gap in STEM. Spatial skills, defined as "the ability to generate, retain, retrieve, and transform well-structured images" (Lohman, 1994, p. 1000), occupy great importance in the engineering field and other STEM occupations. Research shows that, on average, females have weaker spatial skills than males. Several theories exist as to why there are gender differences in spatial skills, ranging from genetic explanations (i.e., spatial ability is transmitted as a recessive characteristic on the X chromosome) to environmental causes (i.e., females have fewer opportunities to play with building materials, such as Legos, which develop spatial ability). Regardless of the origin of gender differences in spatial skills, the good news is that spatial skills can be easily improved (Sorby, 2005). To this end, an innovative spatial skills course at Michigan Technical University was designed to improve the spatial skills of first-year engineering students. Female students tested 6 years after participating in the course evidenced significant improvement in their spatial skills and higher retention in engineering majors compared with women who had not enrolled in the course (Sorby, 2001).

CONCLUSION

Despite the significant progress that has been made in females' representation in STEM, there is clearly much work to be done. Encouraging females to pursue and remain in STEM fields is critical to both individual success and societal progress. As indicated in this chapter, there are a multitude of factors that contribute to the underrepresentation of females in STEM. Despite the long-standing belief that males are innately superior in their mathematics abilities, there is little empirical support for this belief. Instead, societal influences appear to largely explain the gender gap in STEM. The following section provides recommendations for educators to target gender equity with their own students.

RECOMMENDATIONS

As an educator, there are many strategies that you can implement to increase your female students' interest and self-confidence in STEM.

- A good starting place when you are enacting changes in your classroom to increase gender equity is to examine how your interactions with students may differ by gender. You can do this by having an observer record target behaviors based on your interactions with students. For example, the observer may record the number of times a male student versus female student answers a question. Another target behavior is whether the teacher differentially asks high versus lower order questions based on the gender of the student.
- Make a critical examination of the beliefs and expectations that you have of your students. Research shows that others' expectations exert a powerful impact on individuals' achievement (Bachman et al., 2009).
- Encourage your female students to participate in honors or advanced-placement STEM courses. A one-on-one conversation with a teacher who shows faith in a student's ability to handle advanced STEM coursework may be all the encouragement needed to take that particular leap.

- Provide girls with explicit instruction in spatial skills, particularly in introductory engineering courses.
- Encourage female students to participate in mathematics or science summer programs that are specifically targeted at girls. These programs have been shown to increase females' interest and retention in STEM.
- Provide students with exposure to successful male and female role models in STEM. Decorate your classroom with posters of successful males *and* females in STEM careers that are applicable to the subject area of your course.
- Invite female former students who are currently in college studying STEM areas or employed in STEM professions to speak to your class.
- Make connections between what your students are learning in class and the real world. Talk to your students about the range of careers they can have in mathematics and science.

REFERENCES

Bachman, K. R. O., Hebl, M. R., Martinez, L. R., & Rittmayer, A. D. (2009). AWE Literature overview: Girls' experiences in the classroom. *Society of Women Engineers' Assessing Women and Men in Engineering and the Center for the Advancement of Scholarship on Engineering Education.* Retrieved from http://www.engr.psu.edu/awe/misc/ARP_WebPages/classroom.aspx

Beede, D. N., Julian, T. A., Langdon, D., McKittrick, G., Khan, B., & Doms, M. E. (2011). *Women in STEM: A gender gap to innovation.* Economics and Statistics Administration Issue Brief No. 04-11.

Blickenstaff, J. C. (2005). Women and science careers: Leaky pipeline or gender filter? *Gender and Education, 17,* 369–386.

Brotman, J. S., & Moore, F. M. (2008). Girls and science: A review of four themes in the science education literature. *Journal of Research in Science Teaching, 45,* 971–1002.

Burke, R. J., & Mattis, M. C. (2007). *Women and minorities in science, technology, engineering and mathematics: Upping the numbers.* Northampton, MA: Edward Elgar.

Chang, A., Sandhofer, C. M., & Brown, C. S. (2011). Gender biases in early number exposure to preschool-aged children. *Journal of Language and Social Psychology, 30,* 440–450.

Cvencek, D., Greenwald, A. G., & Meltzoff, A. N. (2011). Measuring implicit attitudes of 4-year-olds: The Preschool Association test. *Journal of Experimental Child Psychology, 2,* 187–200.

Eccles, J. S., Freedman-Doan, C., Frome, P., Jacobs, J., & Yoon, K. S. (2000). Gender-role socialization in the family: A longitudinal approach. In T. Eckes & H. M. Trautner (Eds.), *The developmental social psychology of gender* (pp. 333–360). Mahwah, NJ: Erlbaum.

Furnham, A., Reeves, E., & Budhani, S. (2002). Parents think their sons are brighter than their daughters: Sex differences in parental self-estimations and estimations of their children's multiple intelligences. *Journal of Genetic Psychology, 163,* 24–39.

Griffith, A. L. (2010). Persistence of women and minorities in STEM field majors: Is it the school that matters? *Economics of Education Review, 29,* 911–922.

Hill, C., Corbett, C., & St. Rose, A. (2010). *Why so few? Women in science, technology, engineering, and mathematics.* Washington, DC: American Association of University Women.

Kinzie, J. (2007). Women's paths in science: A critical feminist analysis. *New Directions for Institutional Research, 133,* 81–93.

Kurtz, B. C., Rowley, S. J., Harris-Britt, A., & Woods, T. A. (2008). Gender stereotypes about mathematics and science and self-perceptions of ability in late childhood and early adolescence. *Merrill Palmer Quarterly, 54,* 386–409.

Lohman, D. F. (1994). Spatial ability. In R. J. Sternberg (Ed.), *Encyclopedia of intelligence* (Vol. 2, pp. 1000–1007). New York, NY: Macmillan.

Muller, P. A., Stage, F. K., & Kinzie, J. (2001). Science achievement growth trajectories: Understanding factors related to gender and racial-ethnic differences in precollege science achievement. *American Educational Research Journal, 38,* 981–1012.

National Academy of Sciences, National Academy of Engineering, & Institute of Medicine (NAS, NAE, & IOM). (2005). *Rising above the gathering storm: Energizing and employing America for a brighter future.* Washington, DC: National Academies Press.

Schiebinger, L. (2010). *Gender, science, and technology.* United Nations Division for the Advancement of Women, Expert Group Meeting.

Sevo, R. (2009). The talent crisis in science and engineering. *Assessing Women and Men in Engineering.* Retrieved from http://www.engr.psu.edu/AWE/ARPResources.aspx

Sorby, S. A. (2001). A course in spatial visualization and its impact on the retention of women engineering students. *Journal of Women and Minorities in Science and Engineering, 7,* 153–172.

Sorby, S. A. (2005). Assessment of a new and improved course for the development of 3-D spatial skills. *Engineering Design Graphics Journal, 69,* 6–13.

Spelke, E. S. (2005). Sex differences in intrinsic aptitude for mathematics and science? A critical review. *American Psychologist, 60,* 950–958.

Spencer, S. J., Steele, C. M., & Quinn, D. M. (1999). Stereotype threat and women's math performance. *Journal of Experimental Social Psychology, 35,* 4–28.

Stake, J. E. (2006). The critical mediating role of social encouragement for science motivation and confidence among high school girls and boys. *Journal of Applied Social Psychology, 36,* 1017–1045.

Steinke, J. Lapinski, M. K., Crocker, N., Zietsman-Thomas, A., Williams, Y., Evergreen, S. H., & Kuchibhotla, S. (2007). Assessing media influence on middle school aged children's perceptions of women in science using the draw-a-scientist test (DAST). *Science Communication, 29,* 35–64.

Tyson, W., Lee, R., Borman, K. M., & Hanson, M. (2007). Science, technology, engineering, and mathematics (STEM) pathways: High school science and math coursework and postsecondary degree attainment. *Journal of Education for Students Placed At Risk, 12,* 243–270.

Watt, H. M. G., Eccles, J. S., & Durik, A. M. (2006). The leaky mathematics pipeline for girls: A motivational analysis of high school enrollments in Australia and the USA. *Equal Opportunities International, 25,* 642–659.

CHAPTER 11

GEOGRAPHIC DISPARITIES IN SCIENCE ACHIEVEMENT

The Case of Metropolitan St. Louis

Brittni D. Jones
William F. Tate
Washington University in St. Louis

INTRODUCTION

In the United States, the rising demand for human capital in the fields of science, technology, engineering, and mathematics (STEM) continues to increase rapidly (Lacey & Wright, 2009). In addition, STEM occupations rank among those awarding the highest salaries during the beginning career stage (Beede et al., 2011). Results of international comparisons of students' performance in science and mathematics, however, suggest that a substantial proportion of American children and youth, particularly students of color, might not be adequately prepared for STEM-related postsecondary majors and occupations (Gonzales et al., 2009; National Research Council [NRC], 2011). While some scholars have argued that international comparisons are helpful for thinking about federal policy, it is vitally important to understand the nature of science achievement and attainment

Improving Urban Schools, pages 169–189

in the context of urban metropolitan regions (Tate, Jones, Thorne-Wallington, & Hogrebe, 2012). The urban metropolitan region provides a context to understand the geography of opportunity for groups indigenous to a locale. To compete economically, urban regions should prioritize the recruitment of human capital. Failure to provide sufficient resources focused on education, training, and other strategies to support the development of indigenous residents might further exacerbate existing levels of poverty and of related negative social outcomes.

The opportunity to learn science is part of a broader literature referred to as the "geography of opportunity" (Briggs, 2005; Tate, 2008). To evaluate the viability of a metropolitan region to develop economically competitive technology-based industry, scholars must document and analyze the geospatial distribution of science achievement and attainment in the region. In addition, this type of evaluation provides insights into the ability of the citizenry to engage complex scientific questions that have become common in political debates. These dual aims are foundational to discussions of equity in science education. If the distribution of science achievement in a region is uneven, then particular segments of the student population might be ill-prepared to capitalize upon regional economic opportunities and to participate with understanding in a democracy that must engage social problems involving science and health.

This chapter will address two specific objectives. First, we will examine the nature and distribution of science achievement as measured by a state accountability assessment in metropolitan St. Louis. The analysis will focus on fifth- and eighth-grade students in 29 school districts from 2008 through 2011. These grade levels represent foundational learning and developmental pathways for achievement and attainment in high school. Next, the analysis will explore variations in science proficiency and in the demographic composition of school districts. The second major goal is to discuss potential interventions and strategies that might address the geospatial variation in science achievement in metropolitan St. Louis.

We have organized the chapter into three sections. The first will briefly review the role of science and science education in the St. Louis metropolitan economic strategy. We have provided this overview as a reminder that science and technology help to frame opportunity in employment, housing, and education, for example, for indigenous groups. Industrial clusters centered on science and technology represent hubs of employment for communities. Government and private enterprise form regimes to more effectively coordinate options concerning housing and education for targeted initiatives, including emerging industrial centers (Stone, 2005). A discussion of equity in science that fails to capture this reality risks missing the importance of science education for traditionally underserved groups. In the second section, we will review regional science proficiency among students in the fifth and eighth grades. This review includes a description of science proficiency by district and an analysis of demographic variation in science proficiency. To address this variation, we will provide programmatic and policy recommendations in the final section.

METROPOLITAN ST. LOUIS: SCIENCE AS OPPORTUNITY

According to the 2011–2015 Campaign for a Greater St. Louis (St. Louis Regional Chamber & Growth Association [RCGA], 2011), the quality of the regional workforce is the most important issue regarding economic development. The report argued that the lack of a smart, well-educated workforce hinders the ability of a community to attract companies and high-paying jobs to a region. As in similar reports, the document describes strategic steps for improving the regional talent pipeline. We will discuss these strategies later in the chapter.

Before turning to the discussion of these strategies, we offer some insight into relatively recent trends in hiring in the St. Louis market to illustrate the importance of science as a factor related to economic opportunity. The RCGA (2011) documented 74,928 job postings by firms located in the St. Louis metropolitan region in the 120 days between November 2010 and February 2011. Among those with postings, firms in two metropolitan core areas, St. Louis City and St. Louis County, listed 79%.

Nearly half (46%) of all posted opportunities for employment were categorized as computer and mathematical, sales, and healthcare fields. Table 11.1 provides a list of the top 12 opportunities for employment by major occupational category.

TABLE 11.1 Top 12 Job Postings by Firms Located in St. Louis, November 2010–February 2011

Occupation	Total
Computer & Mathematical	14,711
Sales & Related Occupations	10,799
Healthcare Practitioners & Technical	9,243
Other	8,075
Office & Administrative Support	7,910
Management Occupations	7,593
Business & Financial Operations	5,113
Miscellaneous	2,348
Architecture & Engineering	2,320
Food Service	2,149
Transportation	2,144
Life, Physical, and Social Science	1,224

Source: Adapted from RCGA, 2011

Relatively slow population growth in the metropolitan region of St. Louis impedes the task of meeting the changing demand for human capital. In addition, migration flows are not strong in the St. Louis metropolitan region. Weak trends in migration are a primary hindrance and, therefore, ensure that the regional economy may grow slower than will the national economy (RCGA, 2011).

Migration trends are related to the diversity of the talent pool. A diverse talent pool is viewed as a strategic economic advantage for a region. Of the nearly 3 million people living in the 16-county St. Louis MSA, an estimated 77% are White, 18% are Black, 2% are Asian, and 3% are other or two or more races. Approximately 2.5% of regional residents are Hispanic. Note that Hispanics can be of any race. Relative to other emerging hubs of technology, St. Louis is less diverse racially. St. Louis City, Missouri, a county equivalent, is classified within the Midwest and labeled as the only county equivalent outside the South where Blacks surpassed 50% of the total population. More than one researcher has argued that Missouri is a part of the South (Morris & Monroe, 2009; Wells & Crain, 1997). The point here is that patterns of migration and current demographics suggest the need to invest in the science education of the population indigenous to the region.

To create a strategic plan to support regional competitiveness, the RCGA (2011) called for a new framework for the development of talent. The plan consists of three central elements. The first element seeks to increase the level of postsecondary attainment in the region. Expanding upon the first, the second element calls for more individuals with postsecondary education as well as specific propensities and competencies, including people skills, motivation, applied problem-solving skills, and entrepreneurial attitude. The third element envisions developmental learning and career pathways for individuals with the postsecondary credentials and mature competencies to continue accruing skills and understanding required to assume greater responsibility and more demanding leadership roles.

The elements of the St. Louis regional strategic plan for the development of talent appear reasonable. However, the plan fails to assess the status of science proficiency among the students indigenous to the region. This chapter will fill this important void by providing an overview of the attainment of science among students in the metropolitan region of St. Louis. Fifth- and eighth-grade students' performance on the respective Missouri Assessment Program (MAP) standardized science tests from 2008 to 2011 was used to indicate proficiency in science. To understand regional attainment of science, this chapter examines 29 school districts in metropolitan St. Louis. In addition, the chapter will explore variations in the levels of proficiency in science and in the demographic composition of local school districts. Primarily, we aim to emphasize the imperative for leaders of local school districts to ensure equal opportunities for all students to learn and engage in science, regardless of a student's geographic location. Tate (2008) argued that now is the time to engage in research that links the student to technological development in terms of industrial clusters with a full range of developmental outcomes. He contended that this type of analysis would provide insight into the geography of opportunity, which generally grows when left unaddressed.

In metropolitan regions across the country, and in the St. Louis region particularly, geographic disparities have concerned those with a stake in regional economic growth and development (Briggs, 2005; Morris & Monroe, 2009). Cummings (2004) posited that regional disparities in development undermine the ability of community leadership to address spatial mismatch between St. Louis City residents who need jobs and the suburban industrial clusters where job opportunities are being created. A significant amount of the effort related to development is focused on the biotechnology industry (see Table 11.1).[1] In addition, Cummings pointed out the mismatch between the current skill levels of young African Americans seeking jobs and the requirements to compete realistically for new jobs in biotechnology and related industries. He suggested that the reform of workforce development is perceived as a Black issue or a Black program. Thus, the political debate concerning this topic has conformed to the class and racial discourse that have historically split the city and suburban interests in the metropolitan region. The issues of African Americans, workforce development, and related matters associated with educational attainment often produce uncomfortable silences or uninformed explanations. Indeed, both outcomes are unacceptable.

Science Proficiency: A Regional Perspective

Based on data from the 2000 U.S. Census, 29 school districts with the greatest concentration of school-aged children and youth were selected to represent the metropolitan region of St. Louis. These districts exclude those in contiguous Illinois and include districts in Jefferson County, St. Charles County, St. Louis County, and St. Louis City. The data for the 29 school districts were obtained from the website of the Center for the Study of Regional Competitiveness in Science and Technology Database Project (http://maps.wustl.edu/cistl/).

On the MAP science tests, students can achieve one of the following four levels: Below Basic, Basic, Proficient, or Advanced. See Appendix A for a summary describing these levels of proficiency. This chapter relies upon the percentage of students at a combined Proficient/Advanced level for the fifth and eighth grades. For our analysis, we selected these grades because of opportunities to intervene during the learning and developmental process before high school. In addition, examining the science proficiency of students in these grades might more substantively inform discussions concerning improving foundational presecondary opportunities to learn for indigenous groups of students in the region. As defined by the Missouri Department of Elementary and Secondary Education (DESE), students should achieve the minimum level of Proficient to satisfy the Show-Me Standards for science.

Fifth Grade

Figure 11.1 presents the percentages of fifth-grade students at three levels of achievement on the MAP science tests administered from 2008 to 2011. In the graph, the bars combine the Proficient and Advanced levels in the top section. The middle section represents the percentages of students achieving Basic proficiency; the bottom section, Below Basic.

Across the 29 metropolitan districts, the percentages of fifth graders scoring Proficient/Advanced varied widely from 5% to 75%. In 12 districts, the median percentage of students scoring Proficient/Advanced was less than that for the State of Missouri (47%). The percentage of students scoring Below Basic varied from 2 to 46%.

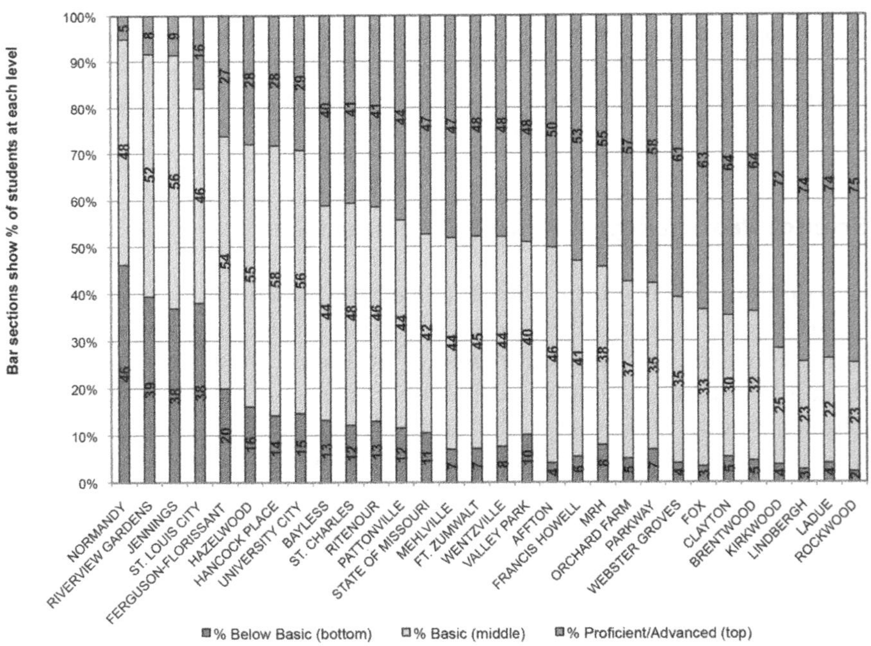

Figure 11.1 Median percentage of students scoring in each level on MAP fifth-grade science test, 2008–2011.

Eighth Grade

Figure 11.2 illustrates the percentages of eighth-grade students at the three levels of proficiency on the MAP science tests administered from 2008 to 2011. The percentages of eighth graders who scored Proficient/Advanced also varied widely across districts (6% to 72%). In 13 districts, the median percentage of students reaching Proficient/Advanced was lower than that for the State of Missouri (47%). In contrast to that for fifth graders (2% to 46%), the range of eighth graders scoring Below Basic varied more widely from 6% to 64%.

To demonstrate the overall performance of students in the 29 districts, Figure 11.3 presents the average percentages of fifth- and eighth-grade students at each level of achievement on the MAP science test. Less than 50% of fifth and eighth graders in local school districts achieved Proficient/Advanced status on the respective MAP science tests from 2008 to 2011. Relative to that for fifth graders (12.7%), a higher percentage of eighth graders (21.0%) scored Below Basic. Administrators and educators within local school districts should be concerned with this recent downward trend.

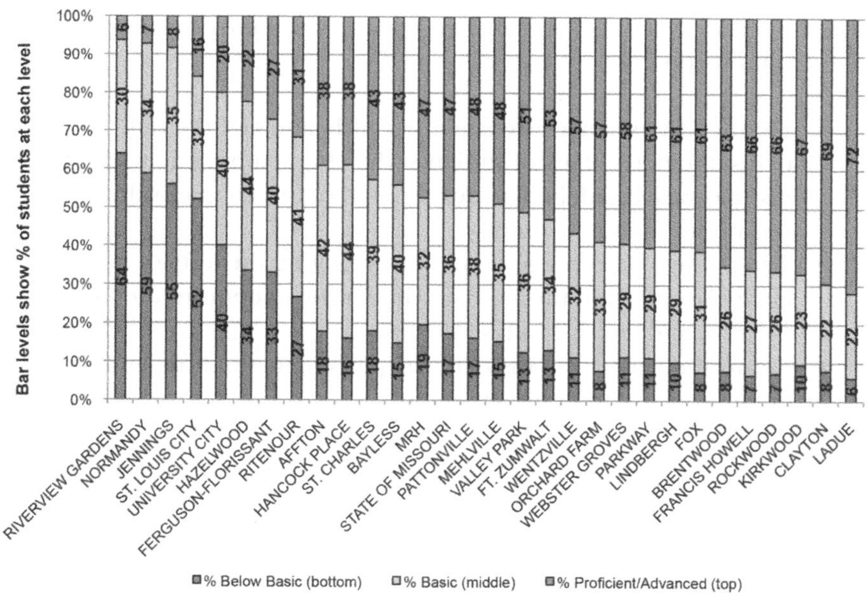

Figure 11.2 Median percentage of students scoring in each level on MAP eighth-grade science test, 2008–2011.

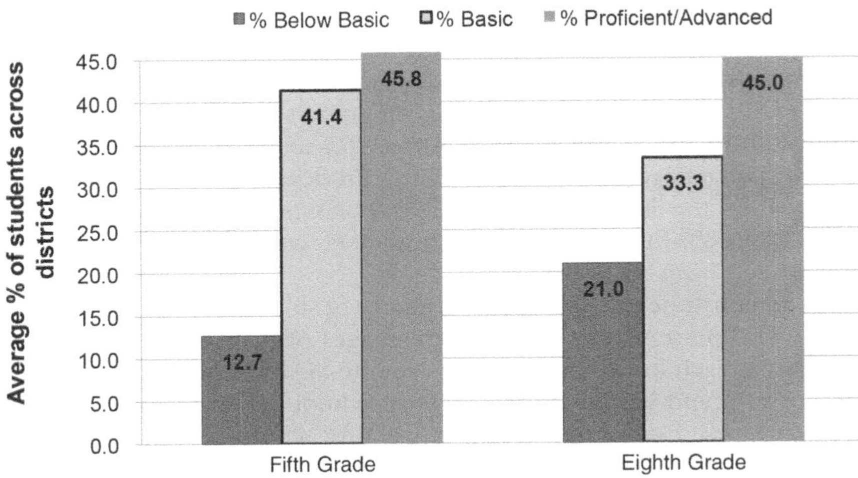

Figure 11.3 Average percentage of students at each level of achievement across 29 metropolitan districts, 2008–2011. The average is calculated from median percentages reported in Figures 11.1 and 11.2.

SUMMARY

The following statements summarize the central points suggested by the data in this section:

- For the MAP science assessments of fifth- and eighth-grade students, the percentage of students who attained proficiency in the 29 districts varied widely. The next section of this chapter will examine this variation more closely.
- The average percentage of students reaching the level of Proficient/Advanced in the area was similar for students in the fifth (45.8%) and eighth grades (45.0%).
- Across the majority of districts, the percentages of fifth- and eighth-grade students attaining Proficient/Advanced status exceeded statewide results.
- Overall, the percentages of eighth-grade students scoring Below Basic exceeded those for fifth-grade students. The average percentage of students scoring Below Basic increased from nearly 13% among those in the fifth grade to 21% for eighth graders.
- In four school districts (Jennings, Normandy, Riverview Gardens, and St. Louis City), less than 20% of fifth- and eighth-grade students achieved proficiency on the respective MAP science tests from 2008

to 2011. More than 50% of eighth-grade students in Jennings, Normandy, Riverview Gardens, and St. Louis City scored Below Basic.

Variation in Science Proficiency Across Districts

To explore more closely the interdistrict variation depicted in the previous bar graphs, this section will use geographic information systems (GIS) mapping. The following two maps (Figures 11.4 and 11.5) display the percentages of fifth- and eighth-grade students who achieved Proficient/Advanced status on the respective MAP science tests in 2011, the most recent year for which data is available. See Appendix B for an alphabetized list of the 29 districts.

Fifth Grade

Figure 11.4 displays the interdistrict variation in proficiency among students in the fifth grade. The map demonstrates a cluster of school districts (white) in which less than nearly one third of students achieved Proficient/Advanced status on the MAP science test in 2011. These districts include St. Louis City and several located to the northwest in St. Louis County. In contrast, at least 50% of students in the majority of school districts in western St. Louis County and in northern St. Charles County (green/brown) attained the level of Proficient/Advanced.

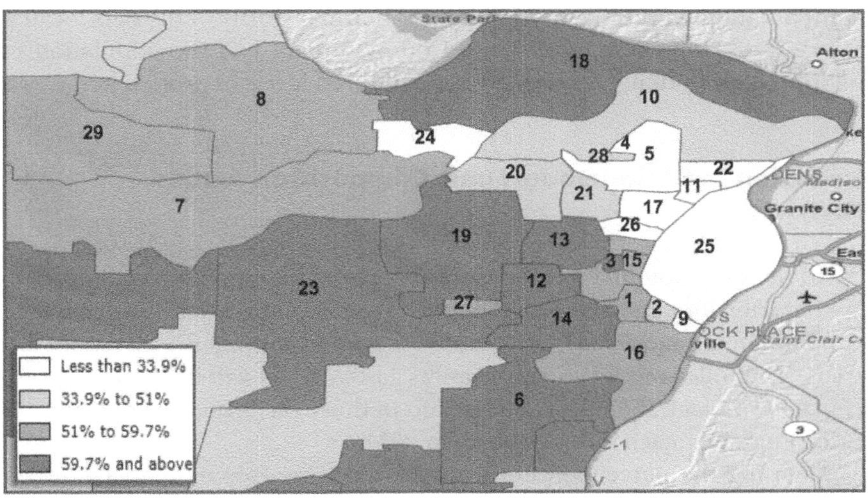

Figure 11.4 Variation in the percentage of fifth-grade students at Proficient/Advanced levels on MAP science test, 2011.

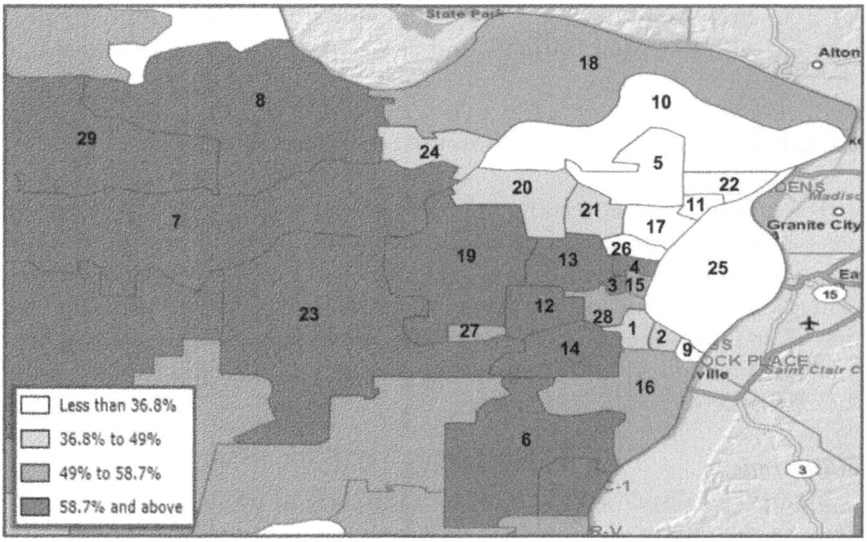

Figure 11.5 Variation in percentage of eighth-grade students at Proficient/Advanced level on MAP science test, 2011.

Eighth Grade

Figure 11.5 presents the interdistrict variation among the percentage of eighth graders who achieved the status of Proficient/Advanced on the MAP science test in 2011. Again, in a cluster that includes St. Louis City and neighboring school districts to the north, less than approximately 37% of students attained proficiency in science. In the majority of districts located in the western portion of St. Louis County, at least half the population of eighth-grade students performed at the desired levels of proficiency.

Variation in Sociodemographic Composition Across Districts

To discover potential relationships between students' sociodemographic characteristics and their attainment of science, this section will explore variation in the racial and socioeconomic composition of the 29 school districts. The following two maps (Figures 11.6 & 11.7) will illustrate the percentages of Black students enrolled and of those receiving free or reduced-priced lunch in each school district in 2011.

Mapping the percentages of students receiving free or reduced-priced lunches indicates the geospatial concentration of low-income and impoverished students in each district. Children from families with incomes at or

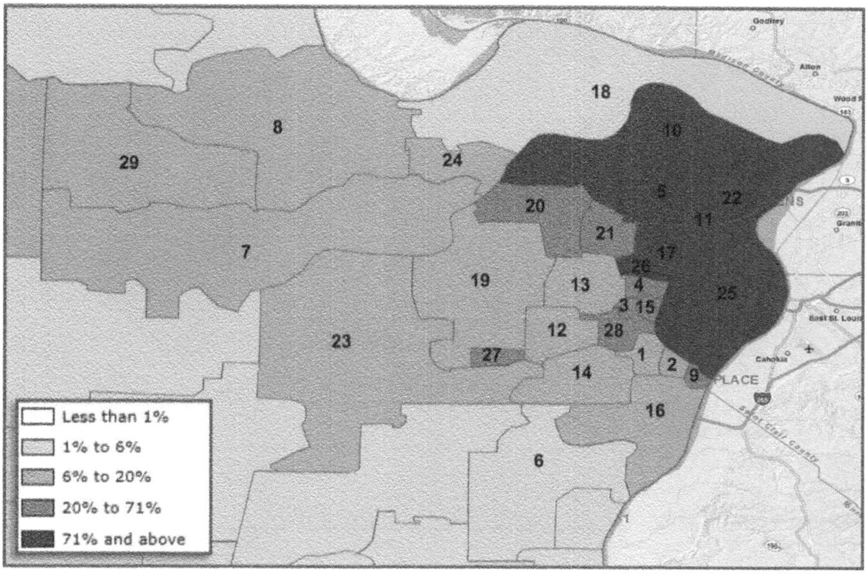

Figure 11.6 Percentage of Black students enrolled in school district, 2011.

below 130% of the federal poverty level ($29,055 for a family of four) are eligible to receive free meals at school. Those with incomes 130% to 185% above the federal poverty level ($29,055–$41,348 for a family of four) are eligible for reduced-priced meals (U.S. Department of Agriculture, 2012).

As demonstrated in Figure 11.6, Black students constitute greater than 70% of the population of students in school districts including and to the immediate northwest of St. Louis City. The enrollment of Black students in Jennings, Normandy, and Riverview Gardens was nearly 100% in 2011. In contrast, Black students compose only 6% to 20% of the total population in school districts in the far western and southern areas of St. Louis County. For instance, in Clayton, Ladue, and Rockwood, districts with some of the highest percentages of Proficient/Advanced students, the proportion of Black students enrolled was 22% or lower in 2011.

Figure 11.7 illustrates the interdistrict variation in the percentage of students who received free or reduced-priced lunch in 2011. In St. Louis City and several other districts located along the border with the State of Illinois, at least 74% of students qualified to receive free or reduced-priced lunches. These districts include Jennings, Normandy, and Riverview Gardens, in which at least 50% of eighth-grade students scored Below Basic on the MAP science tests from 2008 to 2011. In a cluster of districts in the western part of St. Louis County, less than one third of students received free or reduced-priced lunches.

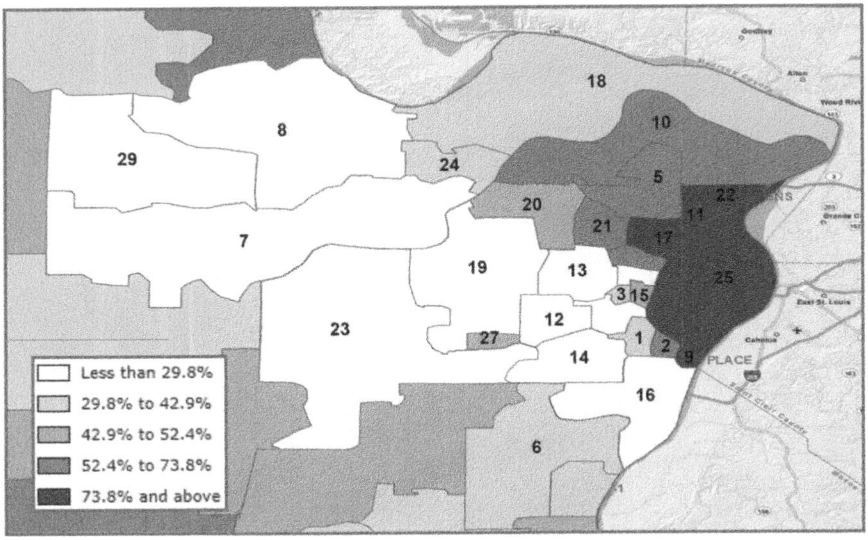

Figure 11.7 Percentage of students receiving free or reduced-priced lunches, 2011.

DISCUSSION

This chapter has attempted to provide some indication of regional attainment in science. The most recent results from the MAP science tests for fifth- and eighth-grade students suggest large variation, by school district, in students' proficiency. By examining this variation and the sociodemographic composition of school districts geospatially, this chapter suggests patterns of lower attainment of science in districts serving heavily concentrated populations of Black students and some of the highest percentages of students eligible for free and reduced-priced meals.

More specifically, St. Louis City and neighboring northwestern school districts, including Jennings, Normandy, and Riverview Gardens, reported some of the lowest percentages of students attaining the status of Proficient/Advanced on the MAP science tests. Relative to those in other regional districts, a higher percentage of students in these four districts scored Below Basic. Moreover, Black students and recipients of subsidized meals composed at least 80% of the total enrollment in the districts.

In other regional districts, such as Clayton, Ladue, and Rockwood, the proportions of Black students enrolled and receiving free or reduced-priced lunches were relatively low. Students in these districts attained some of the highest rates of proficiency on recent MAP science tests. In Ladue, for example, Black students composed only 15% of the population of students in 2011. Approximately 11% of students in the district received free or reduced-priced lunches. This statistic suggests that the school district serves a relatively low percentage of enrolled students from low-income or impoverished families.

As these results demonstrate, some regional school districts with higher proportions of historically underserved and low-achieving students continue to struggle to improve proficiency in science. In contrast, students in more affluent districts in the metropolitan region continue to evince higher levels of proficiency on statewide standardized assessments in science. Therefore, statewide, regional, and district-level educational leaders must evaluate persistent racial and socioeconomic disparities in students' opportunities to learn and engage in science.

Policy and Programmatic Recommendations

This section will offer several recommendations for policies to improve regional science proficiency in metropolitan St. Louis. Particularly, our recommendations will focus on developmental opportunities in youth prior to high school. Other groups and organizations have offered recommendations that deal with shortcomings of the No Child Left Behind (NCLB) Act in the region. For example, since the Riverview Gardens and St. Louis Public Schools districts are unaccredited, several bills introduced recently in the Missouri General Assembly have attempted to ensure access to an adequate education for students across the metropolitan region. The bills have proposed, among other policies, the implementation of vouchers and the creation of a tax-funded scholarship program. Such policies, however, primarily target students within the boundaries of unaccredited school districts. According to evidence provided in this chapter, students in several accredited districts are also not achieving the desired levels of proficiency on statewide standardized science tests. Students in more affluent districts with relatively lower percentages of Black students, however, have consistently demonstrated greater proficiency on the assessments. To improve the opportunity to learn science for low-income and poor Black students in low-performing school districts, we note several areas that represent opportunities for science educators and researchers to contribute to opportunity to learn in urban communities.

Prenatal Process

Extensive research suggests that children of low birth weight experience a greater risk for problems associated with cognitive and academic performance than do their peers of normal birth weight and that the risk for negative outcomes increases as birth weight decreases (Reichman, 2005). In addition, research suggests that the relationship between children of low birth weight (less than 2,000 g or about 4.5 lbs.) and cognitive performance persists through adolescence (Whitaker et al., 2006). Trends of low birth weight have especially challenged members of the community of St. Louis. From 1990 through 2008, the percentage of live births with weight less than 5.5 lbs. has consistently exceeded the national average (Annie Casey Foundation, n.d.). In 2008, the percentage of children classified as low birth weight was 1.42 times greater than the national average.

During pregnancy, smoking cigarettes is the single most well-known cause of low birth weight (see, for example, Anderson, Simonsen, Uldbjerg, Aalkjaer, & Stender, 2009). Feinstein (2011) argued that science education should provide nonscientist adults access to science information to support their decision making about health care, nutrition, reproduction, and other areas associated with their daily lives. Science education researchers and epidemiologists have an opportunity to collaborate on intervention studies focused on prenatal factors associated with cognitive outcomes that are linked to early science learning and understandings. Tate and Striley (2010) noted the need for social scientists in the fields of education and health to work collectively to bridge the divide between the fields. They described the particular need for joint efforts to understand more clearly the mechanisms causing social disparities and to create and evaluate strategies of intervention.

Here, we seek to emphasize the role for professionals of science education in prenatal research and development. The ability to intervene at this stage in the life course of a future student and the family structure has the potential for wide-ranging intergenerational academic effects. Previous discussions regarding opportunity to learn have addressed this recommendation; however, it has suffered in terms of a press to change policy and practice (see, for example, Gordon, 1970; Jackson, 1993).

Birth to Age 4

Despite scholarship that describes the significance of family learning environments and preschool education, members of the science education community have yet to adopt a language that includes parents and families in the process of school reform or as part of research priorities (Weiss, Knapp, Hollweg, & Burrill, 2002). Instead, science reform often uses vague terms and largely empty discourse with respect to family-related interventions and supports. However, a growing body of empirical research focuses on early STEM learning and families and suggests that parents play a critical role in their children's development of conceptual thinking related to science (e.g., Fender & Crowley, 2007).

Despite the substantial potential of the family context as a site for learning STEM content, Cannon and Ginsberg (2008) reported that parents admitted to lacking goals for and knowledge about early mathematics. In addition, parents indicated that early mathematics was taught less often at home than was language-related content, was less interesting to preschoolers, and was less of a personal interest and strength. Cannon and Ginsberg underscored the need for parental interventions that support preschoolers' mathematics learning through daily activities in the home.

We mention mathematics here because it forms part of the foundation of science learning more broadly defined. Fender and Crowley's (2007) research suggests that parental interventions might also improve the learning of science in the setting of the home. We believe that scholars of science education should invest further in this area. One step toward addressing this void is to include, and move beyond, the traditional paradigmatic boundaries of science education activities to other social science fields in which family-level interventions are more normative (e.g., social work, health services, etc.; Corcoran, 2003). We contend that for children up to age 4 and who reside in urban communities, researchers of science education need to examine more closely how family-based mechanisms support understandings of science.

Another opportunity to support positive outcomes related to science achievement may be gained through high-quality urban preschools. We acknowledge a need to document and circulate widely in the urban context the names and locations of high quality preschool providers. Although current scholarship informs an evidence-based framework to support family decision-making related to preschool placement, we remain unconvinced that this research has been sufficiently advertised and implemented in a manner consistent with its importance for science learning (see for example, Hustedt, Barnett, & Friedman, 2010; NICHD Early Child Care Research Network, 2000).

In the context of metropolitan St. Louis, very little research-based information exists that supports parental decision-making related to preschool placement. As is the case in St. Louis, leaders of urban school districts commonly include pre-K initiatives within their plans for strategic change. What is less clear about this effort is how evidence is guiding the design and programming of the strategy and whether or not the strategy is recognized as part of a process to improve science learning and proficiency.

Ages 5–13: Pre-High School Years

During the years preceding high school, students' opportunity to learn science is at risk. In many urban school districts, formal instruction in science is not central, when compared with the relative investment of time and resources provided to reading and mathematics. State accountability testing initiated with NCLB incorporates indicators of ability in reading and mathematics. Thus, school districts have an incentive to prioritize these two areas of study in terms of the allocation of instructional time. Directed by the Center on Education Policy (CEP, 2008), a nationally representative follow-up survey of 349 responding school districts reported that 28% of the school districts indicated since 2001–2002 (the start of the requirements of NCLB) decreasing instructional time for science at the elementary level. Districts that have decreased instructional time for science have done so by an average of 75 minutes per week. In contrast, 58% and 45% of school districts in the survey reported that they have increased instructional time for English language arts (average of 141 minutes per week) and mathematics (average of 89 minutes per week), respectively. A straightforward recommendation, allotting an appropriate amount of instructional time for science education in the pre-high school years can contribute to efforts to increase opportunity to learn science.

Only a few of several potential reforms, these suggestions will require significant cooperation between administrators, counselors, and teachers within local school districts. Each of these recommendations requires careful evaluation of the current standards, curricula, and instructional practices regarding science. As the evidence in this chapter suggests, certain districts will need to consider more robust reforms than will others. In addition, this chapter demonstrates that school districts in the metropolitan region serve students from varying sociodemographic backgrounds. Therefore, when assessing the costs and benefits of these and other reforms, leaders of local districts should do so within their particular contexts with an understanding that their actions will impact the broader region.

Regardless of the direction of local educational leaders, they must cooperate to ensure that students of all races and socioeconomic backgrounds receive substantive opportunities to engage in and learn science. To prepare local children and youth for postsecondary coursework and careers in science-based industries, school districts in metropolitan St. Louis must elevate the value of learning and teaching science. As national and regional projections indicate, those with greater proficiency in science-focused subjects may encounter increased opportunities for employment and economic stability, informed political participation, and a clearer understanding of health and human development.

APPENDIX A
MAP SCIENCE ACHIEVEMENT-LEVEL DESCRIPTORS[2]

Grade 5

Below Basic

Students identify the relationship between mass and force; classify bodies of water; identify weather instruments and their uses; identify characteristics of the solar system; compare amounts/measurements given in a simple format; identify appropriate tools for simple scientific measurements; identify how technological advances may be helpful to humans.
MAP score range: 470–625.

Basic

Students explain the relationship between mass and force; describe how specialized body structures help animals survive; match environments to the plants and animals they support; identify environmental problems and find solutions; determine the appropriate scientific tool and its function in an investigation; determine how technological advances address problems and enhance life.
MAP score range: 626–668.

Proficient

Students describe changes in properties of matter; identify uses of simple machines; explain how work is done; identify forces of magnetism; describe the motion of objects; identify plant parts and their functions; classify vertebrates and invertebrates; classify producers, consumers, or decomposers; predict changes in food chains; identify the effects of human activities on other organisms; describe the Sun as a source of light and heat or the moon as a reflector of light; explain the day/night cycle; interpret data; distinguish between man-made and natural objects; apply problem solving skills to a situation.
MAP score range: 669–691.

Advanced

Students identify energy transformations; predict the effect of heat energy on water; diagram a complete electrical circuit; predict how simple machines affect the force needed to do work; describe the effects of weathering and erosion on Earth's surface; describe relationships in weather data; explain how the Sun's position and the length and position of shadows relate to the time of day; interpret and apply knowledge from a data table; identify appropriate steps and tools in an investigation.
MAP score range: 692–855.

Grade 8

Below Basic

Students identify simple terms related to matter and energy; demonstrate beginning understanding of properties of light and how it travels; identify structures of plants and animals needed for survival; identify levels of organization in multicellular organisms; read simple graphs and make simple data comparisons.
MAP score range: 540–670.

Basic

Students identify an example of a force; demonstrate simple understanding of how traits are passed from one generation to the next; have a basic understanding of climate; identify a simple hypothesis; recognize a trend in a data table; demonstrate some awareness of how various factors influence and are influenced by science and technology.
MAP score range: 671–702.

Proficient

Students classify types of motion; calculate the speed of an object; demonstrate simple understanding of life processes; classify and/or show relationships between organisms; explain how adaptations help organisms survive; explain how species are affected by environmental change; understand and describe a food web; explain rock and fossil evidence of changes in the Earth; explain how Earth's systems interact; draw conclusions from tables or graphs; demonstrate basic understanding of the solar system; recognize the need for, and calculate, averages; use appropriate tools and methods to collect data; describe tools and discoveries that advance scientific knowledge.
MAP score range: 703–734.

Advanced

Students explain the physical and chemical properties of matter; apply knowledge of energy and energy transfer; demonstrate understanding of physical and chemical processes of organisms; evaluate the effects of balanced and unbalanced forces; predict the impact of environmental change in ecosystems; justify how adaptations help organisms survive; demonstrate understanding of the water cycle; compare and contrast weather and climate; explain the cause of seasons on Earth; demonstrate understanding of the solar system; apply the concept of light years; apply awareness of the influence of science and technology in society.
MAP score range: 735–895.

Source: Missouri Department of Elementary and Secondary Education. Retrieved from http://www.dese.mo.gov/divimprove/assess/documents/asmt-gl-gir-spring-2011.pdf

APPENDIX B
29 School Districts Selected to Represent Metropolitan St. Louis

1. Affton
2. Bayless
3. Brentwood
4. Clayton
5. Ferguson-Florissant
6. Fox
7. Francis Howell
8. Ft. Zumwalt
9. Hancock Place
10. Hazelwood
11. Jennings
12. Kirkwood
13. Ladue
14. Lindbergh
15. Maplewood-Richmond Heights (MRH)
16. Mehlville
17. Normandy
18. Orchard Farm
19. Parkway
20. Pattonville
21. Ritenour
22. Riverview Gardens
23. Rockwood
24. St. Charles
25. St. Louis City
26. University City
27. Valley Park
28. Webster Groves
29. Wentzville

NOTES

1. Table 11.1 lists "Healthcare Practitioners & Technical" and "Life, Physical, and Social Science" as separate categories. However, both directly relate to the regional development of biotechnology. In addition, every category on the list is, to some extent, directly or indirectly linked to the expansion of biotechnology and information technology in the region.
2. This information is a direct quotation from the source.

REFERENCES

Anderson, M. R., Simonsen, U., Uldbjerg, N., Aalkjaer, C., & Stender, S. (2009). Smoking cessation early in pregnancy and birth weight, length, head circumference, and endothelial nitric oxide synthase activity in umbilical and chorionic vessels: An observational study of healthy singleton pregnancies. *Circulation, 119,* 765–767.

Annie Casey Foundation. (n.d.). *Kids count data center.* Retrieved from http://datacenter.kidscount.org/data/acrossstates/Default.aspx

Beede, D., Julian, T., Khan, B., Lehrman, R., McKittrick, G., Langdon, D., & Doms, M. (2011, September). Education supports racial and ethnic equality in STEM: Executive Summary. (No. 05-11). *U.S. Department of Commerce, Economics and Statistics Administration.* Retrieved from http://www.esa.doc.gov/sites/default/files/reports/documents/educationsupportsracialandethnicequalityinstem_0.pdf

Briggs, X. (Ed.). (2005). *The geography of opportunity: Race and housing choice in metropolitan America.* Washington, DC: Brookings Institution.

Cannon, J., & Ginsberg, H. P. (2008). "Doing the math": Maternal beliefs about early mathematics versus language learning. *Early Education and Development, 19,* 238–260.

Center on Education Policy (CEP). (2008, February). *NCLB year 5: Instructional time in elementary schools: A closer look at changes for specific subjects.* Washington, DC: Author.

Corcoran, J. (2003). *Clinical applications of evidence-based family interventions.* New York, NY: Oxford University Press.

Cummings, S. (2004). African American entrepreneurship in the St. Louis metropolitan region: Inner city economics and dispersion to the suburbs. In B. Baybeck & E. T. Jones (Eds.), *St. Louis metromorphosis: Past trends and future directions* (pp. 143–170). St. Louis, MO: Missouri Historical Society.

Feinstein, N. (2011). Salvaging science literacy. *Science Education, 95,* 168–185.

Fender, J. G., & Crowley, K. (2007). How parent explanation changes what children learn from everyday thinking. *Journal of Applied Development Psychology, 28,* 189–210.

Gonzales, P., Williams, T., Jocelyn, L., Roey, S., Kastberg, D., & Brenwald, S. (2009). *Highlights from TIMSS 2007: Mathematics and science achievement of U.S. fourth- and eighth-grade students in an international context* (NCES 2009-001 Revised).

Washington, DC: National Center for Education Statistics, Institute of Education Sciences, U.S. Department of Education.

Gordon, E. W. (1970). Education for socially disadvantaged children (Special Issue). *Review of Educational Research, 40*(1), 1–12.

Hustedt, J. T., Barnett, W. S., & Friedman, A. H. (2010, November). *The New Mexico pre-K evaluation: Impacts from the fourth year (2008–2009) of New Mexico's state-funded pre-K program.* New Brunswick, NJ: National Institute for Early Education Research, Rutgers University.

Jackson, S. A. (1993). Opportunity to learn: The health connection. *Journal of Negro Education, 62,* 377–393.

Lacey, T. A., & Wright, B. (2009). Occupational employment projections to 2018. *Monthly Labor Review, 132*(11), 82–123.

Morris, J. E., & Monroe, C. R. (2009). Why study the U.S. South: The nexus of race and place in investigating Black student achievement. *Educational Researcher, 38*(1), 21–36.

National Research Council. (2011). *Successful K–12 STEM education: Identifying effective approaches in science, technology, engineering, and mathematics.* Washington, DC: National Academies Press.

NICHD Early Child Care Research Network. (2000). Characteristics and quality of child care for toddlers and preschoolers. *Applied Developmental Science, 4*(3), 116–135.

Reichman, N. E. (2005). Low birth weight and school readiness. *Future of Children, 15*(1), 91–116. Retrieved from https://www.princeton.edu/futureofchildren/publications/docs/15_01_06.pdf

St. Louis Regional Chamber & Growth Association (RCGA). (2011, June). *Talent: The future of metro St. Louis in the knowledge economy.* St. Louis, MO: Author.

Stone, C. (2005). Looking back to look forward: Reflections on urban regime analysis. *Urban Affairs Review, 40,* 309–341.

Tate, W. F. (2008). "Geography of opportunity": Poverty, place, and educational outcomes. *Educational Researcher, 37*(7), 397–411.

Tate, W. F., Jones, B. D., Thorne-Wallington, E., & Hogrebe, M. C. (2012). Science and the city: Thinking geospatially about opportunity to learn. *Urban Education, 47*(2), 399–433.

Tate, W. F., & Striley, C. (2010, June 23). Epidemiology and education research: Dialoguing about disparities. *Teachers College Record.* Retrieved from http://www.tcrecord.org/Content.asp?ContentId=16036

U.S. Department of Agriculture. (2012). *National school lunch program: Program fact sheet.* Retrieved from http://www.fns.usda.gov/cnd/lunch/AboutLunch/NSLPFactSheet.pdf

Weiss, I. R., Knapp, M. S., Hollweg, K. S., & Burrill, G. (2002). *Investigating the influence of standards: A framework for research in mathematics, science, and technology education.* Washington, DC: National Academies Press.

Wells, A. S., & Crain, R. L. (1997). *Stepping over the color line: African-American students in White suburban schools.* New Haven, CT: Yale University Press.

Whitaker, A. H., Feldman, J. F., Lorenz, J. M., Shen, S., McNicholas, F., Nieto, M., . . . Paneth, N. (2006). Motor and cognitive outcomes in nondisabled low-birth-weight adolescents. *Archives of Pediatrics & Adolescent Medicine, 160,* 1040–1046.

ABOUT THE CONTRIBUTORS

Jason T. Abbitt is associate professor in the Department of Educational Psychology at Miami University. Prior to joining the faculty of Miami University, Dr. Abbitt was a facilitator of professional development workshops for K–12 and higher education faculty focusing on the integration of technology into teaching and was also a designer of web-based learning and collaboration systems. Dr. Abbitt's current research interests focus on the preparation of preservice teachers for technology integration into classroom teaching practices and the integration of emerging technologies into higher education courses and distance learning environments. Dr. Abbitt is the managing editor for the *Journal of Interactive Online Learning*, and his published works appear in the *Journal of Research on Technology in Education*, the *Journal of Digital Learning in Teacher Education*, and the *Association for the Advancement of Computing in Education Journal*, among others. In 2009, Dr. Abbitt received a Faculty Innovator award from the University System of Ohio for his work using technological innovations in his courses for preservice teachers. Dr. Abbitt can be reached by e-mail at abbittjt@muohio.edu.

Brian Boyd is assistant professor of mathematics education at Wright State University (WSU) and is the Founding Principal of the Dayton Regional STEM School (DRSS). Prior to his work at WSU and DRSS, Dr. Boyd worked as a high school mathematics teacher, as well as a middle school principal, in districts near Dayton, Ohio. Dr. Boyd's scholarly interests include the use of inquiry and problem-based mathematics materials, assessing students at higher cognitive-demand levels in mathematics and school leadership. Dr. Boyd can be reached by e-mail at brian.boyd@wright.edu.

Improving Urban Schools, pages 191–197

Mary Margaret Capraro is associate professor in the Department of Teaching, Learning, and Culture at Texas A&M University. She received her PhD from the University of Southern Mississippi. She joined Texas A&M University in 2000 as a clinical professor in mathematics education. Her research interests include teacher knowledge and preparation in mathematics education and student understanding of mathematical concepts. She was previously employed with the Miami-Dade County Schools as both a teacher and assistant principal. She has over 50 peer-reviewed articles and 75 national and international presentations. She is currently Co-PI of Aggie STEM and works extensively with public schools and school districts around Texas, planning mathematics PDs and designing interdisciplinary project-based learning activities. Dr. Capraro can be reached by e-mail at mmcapraro@tamu.edu.

Robert M. Capraro is professor of mathematics education in the Department of Teaching, Learning, and Culture at Texas A&M University and Co-Director of the Aggie STEM Center. He was appointed by the State Board of Education's Expert Review Panel overseeing the mathematics curriculum development. Dr. Capraro's expertise is applied research in school settings and program evaluation, the teacher as change agent for school improvement, and student achievement. He is currently involved in research in four school districts, with more than 20,000 students and 80 teachers. His editorial work includes Associate Editor of the *American Educational Research Journal, School Science and Mathematics,* and *Middle Grades Research Journal,* along with serving on several editorial boards and on the Research Advisory Board for the Associate of Middle Level Education, and a Research Scholar with the Educational Testing Service. He has more than 90 publications and overseen more than $4.5 million in research. Dr.Capraro can be reached by e-mail at rcapraro@tamu.edu.

Trina Davis is assistant professor at Texas A&M University in the Department of Teaching, Learning, and Culture. Dr. Davis was President of the International Society for Technology in Education (ISTE) from 2007 to 2009. She is the Director of Glasscock Island (in *Second Life*) and serves as Co-Principal Investigator of the Knowledge for Algebra Teaching for Equity (KATE) Project, funded by the National Science Foundation. Her research includes investigations related to teaching and learning in online and 3-D virtual environments, as well as statewide school technology studies. Dr. Davis can be reached by e-mail at trinadavis@tamu.edu.

Suzanne Franco is associate professor of statistics and research in the Leadership Studies Department at Wright State University (WSU) and the site leader for WSU's partnership with Ohio Education Research Center (OERC). She also serves as Director of Research and Evaluation at the

Dayton Regional STEM School. Recent research and evaluation work with STEM includes studies around student engagement, college readiness, parent involvement, and inquiry into professional development for STEM teachers. For OERC, Dr. Franco has also served as a PI and Co-PI on research into the Ohio Principal Evaluation System (OPES), Ohio's Student Growth Measures for use in evaluation systems, and ongoing research with the Veteran's Administration Simulation Training for Women's Health Continuing Medical Education (CME) initiative. Dr. Franco can be reached by e-mail at suzanne.franco@wright.edu.

Gerald A. Goldin received his PhD in theoretical physics from Princeton University, coordinated Mathematics Education at the University of Pennsylvania, and Science Education at Northern Illinois University before joining the Rutgers University faculty in mathematics, physics, and education. He has published approximately 200 scholarly articles in these fields, receiving the Humboldt Research Prize for his work in quantum physics. In mathematics education, his work has focused on representations in problem solving and the affective domain in mathematical learning and engagement. He organized and served as the first PI and director of New Jersey's Statewide Systemic Initiative, "Achieving Excellence in Mathematics, Science, and Technology Education," funded from 1993 to 2008 by the National Science Foundation (NSF) and the State of New Jersey. From 2005 to 2010, he was director and PI of "MetroMath: The Center for Mathematics in America's Cities," an NSF-funded center for learning and teaching based at three universities, focusing on mathematics education in urban, low-wealth communities. Dr. Goldin can be reached by e-mail at geraldgoldinbackup@yahoo.com.

Matthew Grushon is the Director of Strategic Initiatives and Special Projects for the College of Education and Human Services at Wright State University (WSU), as well as the WSU Liaison to the Dayton Regional STEM School (DRSS). As a part of these roles, Mr. Grushon serves as the Dayton STEM Hub Coordinator for the Ohio STEM Learning Network. Prior to his work with WSU and DRSS, Mr. Grushon was a practicing attorney. He can be reached at matthew.grushon@wright.edu.

Christina Hamilton is a PhD doctoral candidate at the University of North Texas. Her major is curriculum and instruction with a cognate in mathematics education. Ms. Hamilton's research interest is the use of culturally relevant technologies in the pedagogical practices of mathematics teacher educators. Other research interests include the technological pedagogical content knowledge of teacher educators and the mathematics achievement of students of color. Ms. Hamilton can be reached by e-mail at nsbay0405@gmail.com.

Brittni D. Jones is a Chancellor's Graduate Fellow at Washington University in St. Louis, where she is completing doctoral studies in the Department of Education in arts & sciences. She earned a BA degree in biology with honors (Phi Beta Kappa) from Fisk University. Her research interests include broadly the social contexts of education and the development of adolescents in urban communities. Her more specific areas of interest are the social determinants of the attainment in science, technology, engineering, and mathematics (STEM) and the relationship between STEM attainment and health outcomes. Ms. Jones can be reached by e-mail at jones.brittni.d@wustl.edu.

Meredith Jones is a doctoral student in school psychology at Texas A&M University. She received a bachelor's degree in psychology from Smith College in Northampton, MA. For the past 2 years, she has worked as a graduate research assistant at Aggie STEM. In this role, she has been involved in several research studies and developed gender-equity training for teachers. Her research interests include assessment and intervention for early mathematics difficulties. Meredith Jones can be reached by e-mail at meredithjones88@gmail.com.

Teresa Kennedy is the Director of the International Division of the GLOBE Program, working from the University Corporation for Atmospheric Research (UCAR) Satellite Office located at The University of Texas at Tyler. She is responsible for ensuring that the development and ongoing support of GLOBE's 112 partner countries are fully integrated with GLOBE education and science activities, as well as with programmatic activities related to systems development, communications, marketing, sustainability, and evaluation. Teresa has a PhD in education, with a concentration on international/bilingual science education, and is a Professor of Bilingual/ELL/STEM Education at The University of Texas at Tyler. Her teaching experience includes 8 years at the secondary level, 7 years at the elementary/middle school level, and over 15 years in higher education. She has served as a content-based ESL instructor in Spain and has taught science lessons in classrooms around the world. Dr. Kennedy can be reached by e-mail at tkennedy@uttyler.edu.

Saúl I. Maldonado is a doctoral candidate in education at the University of California, Santa Cruz. Formerly a Northwest Pasadena (CA) teacher and Los Angeles college access program director, his teaching and research interests are centered upon literacy in K–12 STEM and teacher preparation. Mr. Maldonado can be reached by e-mail at simaldon@ucsc.edu.

Eduardo Mosqueda is assistant professor of education at the University of California, Santa Cruz. His research focuses on how school context factors such as course-taking patterns and the segregation of low-income students impact the mathematics achievement of Latino students, particularly English Learners (ELs). He completed his doctoral studies at the Harvard Graduate School of Education, where he was awarded the Spencer Dissertation Fellowship. He has taught both middle and high school mathematics. Dr. Mosqueda can be reached by e-mail at Mosqueda@ucsc.edu.

Michael T. Muzheve is assistant professor of mathematics education in the Department of Mathematics at Texas A&M University–Kingsville. He received an MS degree in mathematics and a PhD degree in curriculum and instruction from Texas A&M University. His research interests in mathematics education are in the areas of mathematics representations and teacher education. Dr. Muzheve can be reached by e-mail at michael.muzheve@tamuk.edu.

Michael R. L. Odell is the Associate Vice President for Sponsored Research and Federal Relations at The University of Texas–Tyler. He also holds the Roosth Chair in education and is a professor of STEM education with a joint appointment in the College of Education and the College of Engineering. He also serves as the Executive Director of the Ingenuity Center, one of seven Texas Education Agency-funded STEM centers. Dr. Odell began his career in education as an Earth science teacher in Irving, Texas (1984–1990). He received his BA in geoscience (1984) and MAT in science education (1989) from the University of Texas at Dallas. He went on to earn his PhD in curriculum and instruction/science education (1993) from Indiana University. He is best known for his work with the NASA Opportunities for Visionary Academics (NOVA) program, the National Space Grant College and Fellowship program, Texas Project Lead the Way (PLTW), and Texas STEM Center Coalition. He has served as a National Space Grant Fellow at NASA Headquarters in Washington, DC (1995–1997) and as a Faculty Fellow at NASA Kennedy Space Center (2003) in Florida. He is currently the President of the National Alliance of State Science and Mathematics Coalitions (NASSMC). Dr. Odell can be reached by e-mail at modell@uttyler.edu.

John Ophus is associate professor of biology and science education at the University of Northern Iowa. He holds advanced degrees in curriculum and instruction as well as entomology from the University of Idaho. His undergraduate degrees are in biology and English from Concordia College (Moorhead, MN). Dr. Ophus is known for his work in the area of project evaluation. He also taught junior high physical and life science before moving into higher education. Dr. Ophus can be reached by e-mail at john.ophus@uni.edu.

Nimisha Patel is associate professor of teacher education at Wright State University, with a focus in educational psychology. Dr. Patel has worked with area STEM schools; she has facilitated focus groups centered on student and teacher perspectives of project-based learning. Additionally, Dr. Patel served as a Co-PI on an Ohio Department of Education–funded College Readiness STEM Demonstration Grant. The research included studies on STEM student engagement, student career development, and leadership. Furthermore, Dr. Patel served as an external evaluator for the HB1 STEM grant awarded to Wright State University in 2010. Dr. Patel is currently conducting research in two areas: parent involvement in STEM and non-STEM education in the rural Midwest; and student engagement in STEM education. Dr. Patel can be reached by e-mail at nimisha.patel@wright.edu.

Frances R. Spielhagen is associate professor of education and Director of the Center for Adolescent Research and Development at Mount Saint Mary College, Newburgh, NY. She is a career educator, with 30 years experience as a middle school and high school teacher before entering the academy in 2002. Dr. Spielhagen is Editor-in-Chief of the *Middle Grades Research Journal* and author of *Unsolved Equations: The Algebra Solution to Mathematics Reform* (Teachers College Press, 2011). Dr. Spielhagen may be reached by e-mail at frances.spielhagen@msmc.edu.

William F. Tate IV is the Edward Mallinckrodt Distinguished University Professor in Arts & Sciences at Washington University in St. Louis. He directs the Center for the Study of Regional Competitiveness in Science and Technology and serves as chair of the Department of Education at the university, where he is affiliated with the Center for Applied Statistics, Institute for Public Health, and Urban Studies. He is a past president of the American Educational Research Association (AERA). Tate is a fellow of the association as well as the recipient of a presidential citation for his contributions to theory and methods associated with research on opportunity to learn and social disparities. He has concentrated his research efforts in three areas: social determinants of mathematics, engineering, technology, and science attainment and disparities; geospatial and epidemiological applications to education, health, and human development; and political economy of urban metropolitan regions. Dr. Tate can be reached by e-mail at wtate@wustl.edu.

Jamaal Young is assistant professor at the University of North Texas. His research interest revolves around the use of technology to develop teacher's content and pedagogical knowledge for teaching mathematics. Other emphases include culturally responsive STEM education and mathematics achievement of children of color. Dr. Young can be reached by e-mail at Jamaal.young@unt.edu.

Jemimah Young focuses her attention on preparing preservice teachers to meet the needs of all students. Her research interest is exploration of alternative cultures in education. Other emphases in her work are culturally responsive education and achievement of children of color. Mrs. Young can be reached by e-mail at jemimah.young@unt.edu.

Printed and bound by CPI Group (UK) Ltd, Croydon, CR0 4YY

15/09/2026

14962370-0002